Humans

Claudio Tuniz • Patrizia Tiberi Vipraio

Humans

An Unauthorized Biography

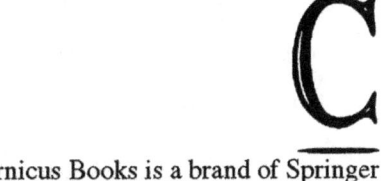

Copernicus Books is a brand of Springer

Claudio Tuniz
International Centre for Theoretical Physics
Trieste, Italy
Centro Fermi, Roma, Italy
University of Wollongong
Australia

Patrizia Tiberi Vipraio
University of Udine (on leave)
Italy

Translated by Juliet Haydock
Sub-edited by Edwin F. Beschler and Marc Beschler

Based on a translation from the original Italian language edition:
Homo sapiens. Una biografia non autorizzata by **Claudio Tuniz, Patrizia Tiberi Vipraio**, Copyright © **Carocci editore 2015**. All Rights Reserved

ISBN 978-3-319-31019-0   ISBN 978-3-319-31021-3   (eBook)
DOI 10.1007/978-3-319-31021-3

Library of Congress Control Number: 2016940134

© Springer International Publishing Switzerland 2016
This work is subject to copyright. All rights are reserved by the Publisher, whether the whole or part of the material is concerned, specifically the rights of translation, reprinting, reuse of illustrations, recitation, broadcasting, reproduction on microfilms or in any other physical way, and transmission or information storage and retrieval, electronic adaptation, computer software, or by similar or dissimilar methodology now known or hereafter developed.
The use of general descriptive names, registered names, trademarks, service marks, etc. in this publication does not imply, even in the absence of a specific statement, that such names are exempt from the relevant protective laws and regulations and therefore free for general use.
The publisher, the authors and the editors are safe to assume that the advice and information in this book are believed to be true and accurate at the date of publication. Neither the publisher nor the authors or the editors give a warranty, express or implied, with respect to the material contained herein or for any errors or omissions that may have been made.

Cover image: © Photo by Joe McNally.

Printed on acid-free paper

This Copernicus imprint is published by Springer Nature
The registered company is Springer International Publishing AG Switzerland

Foreword

Who are we? Where did we come from? How do we relate to other species? How has the history of human evolution on this planet shaped our behaviour and the challenges we now face, individually and collectively?

From the naïve curiosity of early childhood to the more mature reflections of senior citizens, these questions are uncomfortably recurrent. Few people, however, have a clear basis for providing even provisional answers. We tend to conflate what we think we know with what we wish to be so. We need systematic knowledge, based on evidence, rather than prejudices, hunches or faiths.

Serious engagement with understanding our "deep history" takes us into the realms of anthropology, archaeology, natural history, geology, geography and political economy. It is a profoundly scientific endeavour—investigating the known facts, interrogating the evidence and assessing competing interpretations. It is also a profoundly philosophical and ethical undertaking—considering the implications of this knowledge in regard to how we now behave and how we might sensibly change our behaviour or social arrangements.

The questions to be explored are many and varied. What were our origins, going back over millennia? How have we evolved? Should humans be regarded as diverse species? What is the evidence from which we can draw deeper understandings of our human history? And how does modern science help us in identifying and dating that evidence? What do we know about our origins and our evolving relationships with the natural world, including other species (many of which we are responsible for making extinct)? What about the significance in this story of different types of food? Of the capacity to make tools? Of clothing? Of war? Of burial procedures? Or sundry other social arrangements?

It is rare to find authors with the capacity to deal with such a mind-boggling breadth of concerns. But here is the combined work of a physical scientist and a political economist, jointly seeking to clarify the issues in straightforward language for a broad public audience. It is a bold undertaking and, in my judgement, superbly executed. It is tempting to draw a parallel with other previous "popular" books on scientific topics, such as *A Brief History of Time* by Stephen Hawking and *The ABC of Relativity* by Bertrand Russell, setting out to explain scientific complexity with a minimum of fuss.

The book does not presume any significant prior knowledge: an intelligent interest is sufficient. It is recurrently thought-provoking, prompting deep reflection on our origins and nature, while also weaving interesting (and often wry and witty) reflections on contemporary social issues into the historical and scientific narrative. It is clearly structured, progressing step by step through the various interrelated aspects of the topic. The shift in the latter parts towards consideration of cultural and economic themes leads to profoundly challenging conclusions at the end.

Reading this book illustrates the inherently exploratory nature of scientific inquiry and its essentially open-ended character. It is like being invited on a great journey. Better still, two great intersecting journeys—one following the evolution of our species and the other exploring how scientists have sought to analyse that evolution.

Now is a great time to consider where the latter journey has been taking us. Scientific studies of our species' origins have made substantial advances during the last decade. At least half a dozen significant ancestors have been unearthed. New scientific methods for studying human remains have been invented, from genomics to 3D radiation imaging. There has been detailed reconstruction of the landscape in which we evolved, using Antarctic ice, for example. The result of these developments is that we now have a better understanding not only of our anatomical evolution but also of our behavioural evolution.

Ultimately, the important issue is what we do with this knowledge in a difficult and dangerous world. Major economic, social, environmental and political dilemmas confront us. How are we to live in harmony with each other, despite recurrent conflicts associated with differences of class, ethnicity, national identity and religious faith? Can we replace consumerist capitalism, which is so obviously unfulfilling and unsustainable, with economic arrangements that do not depend on the pursuit of endless growth and rapacious relationships to nature? How can we avert the awesome consequences of climate change? Most generally, how are we to create a more balanced relationship with each other, other species and the physical environment? The authors of this book do not claim to have all the answers, but they pose the big questions and they give us knowledge of the historical context within which we can better understand the challenges and possibilities.

Should we be optimistic? The authors emphasise that it is the capacity for symbolic thought that most clearly distinguishes modern humans. This capacity has evolved as the basis for our humanity, providing the scope for behaviours that have enormous potential—whether for public benefit or for social and environmental harm. It gives us the ability to see "the big picture", beyond a merely egocentric perspective, and to consider alternative courses of action for dealing with current challenges. It gives us the capacity to plan, to change direction and to redirect our energies for different social purposes. These are, indeed, prerequisites for acquiring knowledge and using it in fruitful ways. The book is clearly written in that progressive spirit.

Reading this slim volume could change the way you see the evolution of humankind and the challenges we all now face—and even change your personal life course. Journey on...

University of Sydney Frank Stilwell
Sydney, NSW
Australia

Acknowledgements

This little book is the result of many years of discussions with both scholars and non-specialists about the many facets of human evolution and the interplay between natural and social sciences in unveiling our deep past: a no-man's land in the current scientific discourse. It is a work of synthesis. We, therefore, owe a great deal to the many brilliant scholars upon whose research and ideas it draws. It is also an attempt to make recent discoveries accessible to all. We are, thus, grateful to the many comments of the general public and the high school students to whom these ideas were introduced, in various presentations of an early version of this book. In addition, we would like to thank a number of friends and colleagues who were extremely generous in commenting and discussing with us the first draft of it. Our sincere thanks go, in particular, to Colin Groves for his precious advice and remarks, to Frank Stilwell for his thorough comments and enthusiastic support, to Peter Cory for his keen and detailed observations and finally, to Alessandro Tuniz for his merciless criticisms. Needless to say, any errors or imprecisions that remain are all ours.

Contents

1	**History, Prehistory and Deep Time**................................	1
	1.1 Our Viewpoint..	3
2	**Genesis**..	7
	2.1 The Timescales of Evolution............................	9
3	**The *Star Wars* Cantina**...	13
	3.1 Children of Eve...	14
	3.2 Human Dispersions...	15
	3.3 "Open Family" Experiments.............................	16
	3.4 More "Open Family" Experiments and Their Consequences...	20
4	**The Apes and Us**...	23
	4.1 The First Scandals...	24
	4.2 Planet of the Apes...	25
	4.3 The Human Planet...	26
5	**The Quest for Fire**..	29
	5.1 From Scavengers to Opportunists and Hunters............	29
	5.2 Climate and Evolution.......................................	31
	5.3 Weapons of Mass Destruction........................	33
	5.4 The Disappearance of the Neanderthals: Many Clues, Little Evidence..	35
6	**The Naked Ape**...	37
	6.1 The Loss of Hair: A Few Facts and Many Hypotheses......	38
	6.2 Clothing, Roles and Images.............................	40
7	**Lucy and the Other Ladies**..	43
	7.1 The Obstetric Dilemma.....................................	44
	7.2 Turning a Disadvantage into an Advantage..............	46
	7.3 Teeth: The Black Box of Our Lives................	47
	7.4 "Multimedia" Ceremonies................................	48

8	**Menus of the Past**	51
	8.1 Ritual Food	53
	8.2 Vegetarians or Carnivores? Omnivores!	54
	8.3 Us and Them: Cain's Diet	56
9	**Ancient Ills and Ancient Remedies**	61
	9.1 Diseases and Treatments	63
	9.2 Placebo Effect	64
	9.3 Migrations and Contagions	65
	9.4 Self-inflicted Diseases	65
10	**The Hominin Lifestyle**	67
	10.1 Growing Up Too Quickly?	67
	10.2 Art and Entertainment	68
	10.3 Grandparents and Grandchildren	69
	10.4 Monogamy or Polygamy?	70
	10.5 Trust, Gossip and Shared Beliefs	71
11	**The Dearly Departed of the Pleistocene**	75
	11.1 Ceremonies	75
	11.2 The First Hierarchical Societies	77
	11.3 Neanderthal Burials	78
12	**Brain Readers**	79
	12.1 Inside the "Grey Box"	79
	12.2 Mind and Brain in Deep Time	80
	12.3 Building Up the Human Mind	82
	12.4 Thinking About Thought	83
13	**All Power to Imagination**	85
	13.1 New Realities	86
	13.2 Inside the Black Box of Symbolic Thought	89
	13.3 Contemporary Examples	90
	13.4 Possible Explanations	92
	13.5 Symbolic Thought and Cultures	93
	13.6 Excesses of Representation	94
	13.7 A Recap	95
14	**Primordial Economy**	97
	14.1 *Homo Economicus*	97
	14.2 The Origins	98
	14.3 First Evidence	99
	14.4 The Concept of Money	100
	14.5 Real Economy	101
	14.6 Technology, Population and Climate	103

14.7	Specialisation and Generation of Material Wealth	104
14.8	Counter-Evidence	105
14.9	Private Accumulation of Wealth	106

15 (In)Conclusive Remarks ... 109

Further Readings ... 115

Chapter 1
History, Prehistory and Deep Time

Akpa le tome gake menya tsi fe vevie nyenyeo
(A fish is the last to acknowledge the existence of water)
African proverb

Until relatively recently, we knew little about our deep past. History books began with the great civilisations: Egyptian, Phoenician, Hittite and Assyro-Babylonian. Everything that happened before was shrouded in mystery. We used the term Stone Age to describe a long period when we made very little progress in our ability to survive and had very primitive tools for hunting and gathering the gifts of nature. This period was followed by the Copper, Bronze and Iron Ages, all named after the various metals we were able to extract from the rocks and work with the aid of fire. These advances enabled us to equip ourselves with increasingly complex tools.

The development of agriculture and stock farming allowed us to accumulate many surplus resources, which we eventually began to trade. Lastly, we built great monumental works, some of which have lasted to our own day. Then, there were the wars, which the new metal weapons made increasingly bloody: history's main event. Yet we did not know how far we could trace back our origins as human beings. All we could do was place our trust in sacred texts, which told us different stories depending on our religion.

Over the past 150 years, the scientific method has enabled us to shed light on many aspects of our deep history: students of human science refer to this as "prehistory", and it is usually the exclusive preserve of anthropologists and archaeologists. It is generally agreed that it began with the Palaeolithic (a period that occurred approximately between 2.5 million and 10,000 years ago). It continued through the Mesolithic and Neolithic and ended with the first written records. This means that prehistory lasted until about 5500 years ago, at least in the Middle East where the Sumerians were the first to develop a system of writing, thus marking the beginning of "history".

Nineteenth-century European academics were responsible for introducing these classifications and their respective time-scales. These inevitably vary for different human peoples, because each began to record its history at different times, if at all. For this reason, such classifications do not seem very useful for our purposes.

According to these definitions, when the English captain James Cook arrived in Australia in 1770, the Aborigines were still living in the deepest prehistory. They had not felt any need to develop writing, or indeed to invent the wheel or introduce any of the other innovations that we consider essential for "civilisation". Without a civilisation, they were not considered to be worthy of a history. Worse still, they were not even entitled to their own land, given that the continent they had inhabited for more than 50,000 years was immediately defined as *terra nullius* and granted to the British Crown.

The Aborigines actually did hand down their knowledge, but only orally, through a rich tradition of songs, music and dance. Their essential needs were also largely met, leaving room for a cultural life based on a wealth of legends that was deeply spiritual and survives to some extent even today. Their accumulated knowledge—partly known to us and partly lost—is now of great interest to contemporary medicine and other disciplines. They were hunter-gatherers, not prehistoric people. They were not in any way inferior or uncivilised.

Fig. 1.1 The first *Homo sapiens* arrive in Australia, 50,000 years ago. *Source*: Drawing by Tullio Perentin, ZOIC, Trieste

Many other indigenous peoples discovered by Europeans travelling the world during the colonial voyages of discovery a few centuries ago found themselves in a similar situation: deprived of their history only because they had no written record. And if they had a civilisation, as in Central and South America, this was soon swept away by the advent of a "superior" civilisation: the European. Sometimes things got even worse, if we think of the treatment suffered by the African slaves who were deported to the American continent for approximately three centuries of the previous millennium. They were deprived not only of their families, their land and their history but even of their membership in humankind. There is no other explanation for the way the slaves were equated with mere beasts of burden by people who called themselves Christians.

If the term prehistory must continue to be used, it must be acknowledged that this does not have and cannot have any specific time-scale except from a Eurocentric viewpoint, i.e. one that uses the history of Europe and its surrounding areas as a benchmark for the history of all other peoples. We are unwilling to go along with this methodological approach, even though it continues to permeate the debate over our origins.

1.1 Our Viewpoint

In our narrative, we will abandon the term prehistory and will talk about human history in a broad sense, using it to describe not only our own history but also that of the other species—human and proto-human—who came before us. We will often use the colloquial term "our ancestors" to describe those from whom we have descended directly, as well as the broad and heterogeneous family of hominins that broke off from the evolutionary line of the chimpanzees between 7 and 6 million years ago. Some 20 different hominin species have been identified to date. The group is much more limited than that of the hominids, because the latter also includes the other great apes; but it is larger than the group of humans in the strictest sense (belonging to the *Homo* genus, which only includes about 10 species as far as we are aware).

Contrary to popular belief, evolution does not take place in a linear manner but unfolds through an intricate bush that can culminate in many barren branches. This makes it difficult to identify our ancestors appropriately, because we are as yet unaware of all the variants that came before us. When we speak about evolutionary lineages, we are not merely referring to the main lineage but also to all the other branches.

Taking a look at the extended hominin family, we believe we now have a clearer idea of certain steps that are essential for identifying what makes us human. We are even able to understand what marked the difference between humans in the recent past, when at least four species of *Homo* were living simultaneously on our planet. The evolutionary story we are about to relate begins with the assumption of an upright posture, followed by an increase in brain volume, and ultimately the advent of symbolic thought, in other words, the ability to generate symbols, convey ideas and think in hypothetical terms.

Scientific results garnered in recent years have begun to assemble a very specific picture of all aspects of our origins. We can now go back in time to the start of the universe we inhabit, when our planet was formed and when life on Earth began. We are aware of the laws that underpin evolution and changes in living beings, as well as the ties that exist and have existed between different forms of life. We also know when and where we first arose as a species, although many things remain obscure.

The continuing debate is largely conducted among academics using terms that are relatively opaque to non-experts. The methodologies are known only to insiders and the findings are, quite rightly, subject to debate over their credibility, as well as their potential implications. When the findings are eventually reported in the press, they are turned into sensational headlines that miss many crucial points.

We also have to deal with all the preconceptions that permeate our culture; besides, our acquired knowledge all too often loses its form and meaning when made public. This situation leaves room for a series of very tough if not altogether impossible debates. One good example is that between science and religion, two conceptual frameworks that exist on different and barely comparable planes: the former is based on systematic doubt, the second on unconditional (if dogmatic) faith. We do not intend to go into this topic here.

Instead, we will try to set out the facts as they have been established on the basis of scientific investigation. Where there is still no explanation for these facts, we will merely attempt to provide some possible interpretations. We will also try to use language that is accessible, even if this entails less than fully comprehensive accuracy: this is something that scientists generally abhor, and we apologise to them in advance. Our aim is to write a human biography intended for curious non-experts, rather than academics. The technical details as to how we arrived at these results will be given but will be minimal. The extensive literature on these findings is easily available using the normal Internet search engines. Above all, we will recount the history of our origins through the daily lives of our ancestors, the difficulties they faced, their migrations and their most dramatic moments, which, in some cases, brought them to the threshold of extinction. In the concluding sections, we will attempt to distil the essence of our nature.

We will in this way debunk many myths that continue to flavour the discussion about who we are, where we come from and where we are going. What emerges is an adventurous, entertaining and sometimes dramatic picture. The discovery of new human species that lived on this planet with us or came before us is not a recent event. We were already aware of the existence of the Neanderthals in Eurasia, but we leapt to the conclusion that they were a species "less human" than we are. Until a few decades ago, all depictions of Neanderthals described crude and primitive beings. Nothing could have been further from the truth. In recent years, we have discovered more human species about which we still know little but who have nevertheless provided us with surprising information. Perhaps we will discover others, now that we have the means to do so.

In the studies of our origins, palaeoanthropology draws together many disciplines: medicine, neurology, physics, biology, genetics, chemistry, geology, archaeology and even engineering and computer science. Its theories or proposed interpretations are not therefore based on vague clues but on hard facts, dates and evidence.

We can also consider our history in broader terms by introducing human and social sciences, such as psychology, sociology, political economy and demographics. When we do this, we can see a more complex picture emerging than one based simply on our intellectual superiority over all other species. If we cross-match the available data and their interpretations according to individual subject areas, we can put together a general proposed interpretation to explain what made us, at some point, so different from the others, a quality that is right under our noses every day: this difference was almost certainly determined by our ability to imagine worlds that are different from the one we inhabit. The next steps were to represent them, acknowledge them and ultimately broadcast them. This marked the advent of a new and different trait: one that can be traced back to symbolic thought and then into the different cultures according to the time and space in which it could manifest itself.

This ability allowed us to create another reality in our minds that we initially began to reproduce on the walls of our caves or in the objects we were able to fabricate with the materials around us, culminating in our ability to convey it through a system of interpersonal communication. It was something we wished to depict through painting, sculpture, music and, in time, literature, audio-visual arts and so on. Even scientific thought is the outcome of our ambition to establish whether what we originally imagined is actually borne out by a closer analysis of the observed reality. Our ability to imagine also assumed a social and economic dimension, expressing itself in new rules of cohabitation and new technologies, before expanding through the exchange of goods and services. This allowed us to make life easier for ourselves, at least up to a certain point.

If we look at our history as hominins, it can be seen that this trait was already present in embryonic form very long ago. It is illustrated through our understanding that we could better satisfy our needs by adopting tools and resources that we were not endowed with by virtue of our body shapes but that we could gather in nature, moulding and utilising them for our purposes. This point marked the first break between what we are, from an anatomical viewpoint, and what we are able to do if we increase our resources by creating special tools to help us survive. It was the first distinction between *being* and *having*, a theme that has persisted through our own times, reflecting the advent and dominant traits of our evolutionary line. These traits are apparently associated with a sideways expansion in the brain's parietal lobes, which are responsible for coordinating our relationships with tools and their use.

This artificial enhancement, which turned us into humans, nevertheless remained limited for a long time to the mastery of fire and the use of tools requiring minimal working. Small, incremental innovations were applied to natural objects and to their assembly for some two million years, albeit gathering pace at the end of the period.

The radical innovation that grew out of this ancient perspective is represented by symbolic thought. This trait is associated with the individuals we refer to as "modern humans", who also expanded the upper part of their parietal lobes. This event further enhanced our cognitive abilities. One of the first expressions of symbolic thought was the representation of reality through images that we created ourselves. This attribute has now led us to virtual reality: worlds that exist only in our imagination but become real by virtue of our ability to conceive them and communicate them.

As we go through the essential steps in our history, a clear picture of what turned us first into humans and then into modern humans seems to emerge. Even before the great civilisations, our diversity was first translated into the conception of a reality that was slightly different from what we observe in nature, and then to the construction of a reality that was increasingly free from what we can sense and perceive, because it was the result of our thought alone. When it became possible to represent this new reality, it took on a life of its own, leading us to generate worlds that we can inhabit only with our minds. As we began to form larger societies, we gave up our individual independence, but in exchange, we were able to benefit from more goods and services. In order to gain access to these, we had to construct rules and ensure they were followed. We also had to create ties between remote and unknown people. We had to generate increasingly complex cultures as a basis for individual and collective forms of behaviour. This turned us into a social body.

One of the results of this process was our will to control nature and bend it to our purposes, giving rise to the geological era that some refer to as the Anthropocene, in other words, the period when we started to have a detectable impact on the environment. Even though many believe that our global footprint began only with the latest industrial revolution, we will see that its roots go much deeper, dating back to more than 50,000 years ago, when our arrival coincided with the first global extinctions and a change in entire ecosystems.

We must nevertheless go very far back in time to establish the roadmap and boundaries of our long evolutionary pathway.

Chapter 2
Genesis

When we are children, as soon as we begin to reason, the first things we wonder are: Where do we come from? Why are we here? How long have we been here? These simple questions about our origins are soon followed by others concerning the reason for our existence and its purpose. The latter questions are clearly the province of religion and moral philosophy. Yet, while speculation over the time and manner of our origin was once governed by the absolute authority of the sacred texts, such authority has for some time been threatened by a new approach: that of science, which is now capable of giving us specific answers not only as to when life on Earth began, but also when the formation of our planet, our galaxy and the universe in which we live occurred.

The Bible seems to provide, name by name, an accurate list of all the descendants of Adam and Eve. This very precise and detailed information persuaded the Irish Archbishop James Ussher, in 1650, that man was made 6 days after the creation of the world, which took place on the evening of October 22, 4004 BC. This is the pillar on which present-day Creationism stands. This current of thought is relatively popular, even in the most advanced countries. Its most recent manifestation is known as Intelligent Design. Whereas Creationism interprets Genesis in a literal sense, attributing a scientific aspect to the Bible story, Intelligent Design accepts evolution but argues that it takes place through the external intervention of a higher Being.

Scientists did not begin to consider the possibility that Earth and its living creatures had a much more ancient history until a couple of centuries ago. Similar thoughts had hatched in the minds of naturalists of earlier centuries, including Leonardo da Vinci. These ideas were based on methods of investigation that were already being used back in the fifth and sixth centuries BC by the Greek pre-Socratic philosophers, although they would not be permitted to be expressed freely for many centuries. During the final decades of the eighteenth century, James Hutton, who is considered the father of the geological sciences, began to argue that our planet could not have been moulded into its current form in such a short time. Geological processes are slow and act over a very long period. According to Hutton, who argued on the basis of the evidence then available, in the geological history of our planet, "we find no vestige of a beginning, no prospect of an end" (*Theory of the Earth*, 1788, p. 304).

Paradoxically, the information provided by the scientist actually seemed to be much less precise than that supplied by the Archbishop. But science needs to proceed at its own rate: in order to find more reliable information, we must first shed doubt on the information we already possess. Only doubts and inconsistencies give rise to new knowledge, certainly not dogmas or the presumed certainties that we use to contain our anxiety about what we do not know. In actual fact, it was not until over a century later that physicists found new "clocks" for measuring deep time. These were based on the instability of certain radioactive atoms present in nature, including uranium-238, potassium-40, carbon-14 and many others.

In the meantime, the need emerged to consider a longer timescale not only in regard to an explanation of the Earth's origin but also of the origin of life on the planet. Half a century after Hutton, Charles Darwin focused on the latter question; he also became persuaded that the Earth's origin must have been much more ancient than believed. It must have been sufficiently remote in time to allow for the evolution of life into "endless forms most beautiful and most wonderful" (*On the Origin of Species*, 1859, p. 490). Surely, the few millennia of the Archbishop were not enough. Darwin's work described the mechanisms that led to the establishment of all known plant and animal species, emphasising their great variety and their adaptation to different habitats.

Darwin ultimately found himself drawing up a theory about the origin of life that was very much at odds with that claimed by the Anglican Church, particularly once he encroached on the sensitive area of the origins of man. His wife Emma was a fervent Christian and he supported his local church, even though he was not a regular churchgoer. At the time, Darwin felt he would not be able to tackle the question head-on and shelved the matter for later examination. He actually concluded his work with a sentence that seems premonitory: "Light will be thrown on the origin of man and his history" (ivi, p. 488). That was his final word, at least for the time being. It was only in 1871 that he dared to publish his ideas on this hot subject in *The Descent of Man*.

Naturalists immediately grasped the implications of his theory of evolution as to the explanation for human genesis, despite the distorted way in which it was presented. Darwin's thought was interpreted as a claim that man descended from present-day apes, which was a mere caricature of his ideas. The spread of Darwin's ideas, variously misinterpreted, deeply disturbed most believers. Many newspapers of the day published a picture of Darwin's head grafted onto an ape's body. It is said that immediately after hearing this theory, the wife of the Anglican Bishop of Worcester reportedly uttered the famous phrase: "Descended from the apes! My dear, we will hope it is not true. But if it is, let us pray that it may not become generally known". Whether it is true or not, this anecdote nevertheless gives us an idea of contemporary Victorian attitudes. Not until 2008 did the Anglican Church officially apologise to Darwin (who was, in any case, buried in Westminster Abbey) for its original misunderstanding of his theory of evolution. Nevertheless, many representatives of that church still call, some quite recently, for Creationism to be taught in schools, particularly as part of the scientific curriculum, in the version known as Intelligent Design.

2.1 The Timescales of Evolution

Nowadays, the origins of the world can be traced back in great detail to the very beginning. We know that our universe was born with the Big Bang 13.8 billion years ago, when all energy and mass were concentrated in a single point. Astronomical observations, using such means as the European Space Agency Planck satellite, have made it possible to reconstruct its expansion and cooling. Satellite detectors have delivered an accurate map of the fossil radiation produced at the origins of the universe (380,000 years after the Big Bang, to be exact). When the Big Bang took place, time, space and matter took form. Matter then evolved into a multiplicity of particles which, in turn, condensed first in an inanimate form and then, at least on Earth, into life.

Large, extremely dense stars formed through the force of gravity. The nuclear reactions that took place inside of them generated all of the elements that comprise what we think of as ordinary matter: carbon, nitrogen, oxygen, silicon, magnesium, and iron. Some stars exploded and transformed into supernovae, from which the heavier elements, all the way up to uranium, were generated and spread across the galaxy. Over time, these elements became part of the interstellar dust, which formed the first solid bodies through the process of accretion. Our solar system, with all its planets, thus developed from the ashes of stars that exploded during the early years of our galaxy.

The atoms of everything that exists on our planet—from rocks to air, from water to plants and animals—were, therefore, once contained in the core of large stars. Hence, this also applies to the atoms that make up our bodies, which have, through evolution, temporarily assumed the "form" of our species and will continue to move from form to form through the life-cycle, only to return to the universe as dust in the end. Ashes to ashes and dust to dust: the Bible got it right this time.

The latest developments in theoretical physics suggest the existence of many universes that cannot communicate with one another. It also seems that the atoms making up our particular universe—from our bodies to the air we breathe, from the planets to the galaxies—account for only 5 % of the total matter and energy existing in the cosmos. All the rest is part of a form of matter (and energy) that is defined as "dark" and that physicists are actively investigating. As our knowledge grows, the awareness of our ignorance grows as well—yet, this seems to increase our curiosity rather than hold us back.

Going back to our original question, do we at least know when Earth was actually born? Yes, but we have only known since the mid-1950s. The Earth's date of birth was established by measuring lead isotopes produced in rocks through natural uranium radioactivity, isotopes that are generated at a rate known to us (half of the atoms of uranium-238 present in the rock decay in approximately 4.5 billion years). Isotopes are atoms of the same element with a different mass; they are described using numbers that refer to the weight of the corresponding atom. In this case, the atoms in question are lead-206 and lead-207. We can think of this decay process as a "nuclear hourglass", in which the original atoms of uranium in the top bulb turn into atoms of lead that are deposited over time into the bottom bulb. The final number of lead atoms evaluated using special atom counters (known as mass spectrometers) makes it possible to determine the time that has elapsed since the rock was formed with great accuracy.

These counters became available towards the middle of the previous century, after significant investment in a type of research conducted for a very different purpose, namely the construction of the first nuclear bombs. It was ascertained, without a shadow of a doubt, that our planet formed 4.55 billion years ago.

So what about the origin of life? Through the analysis of certain minerals present in the rocks of Western Australia, which have a known chronology, we can establish that our evolutionary history began with the first single-celled organisms, which appeared on Earth over 4 billion years ago. All the bacteria, animal and plant species we know evolved from them.

We have also discovered through genetic studies that our evolutionary line broke off from that of chimpanzees approximately 6 million years ago, when we developed, among other things, a particular gene, ARHGAP11B. According to researchers at the Max Planck Institute in Dresden, this event, which was responsible for the growth of the cerebral cortex, marked a critical difference between us and the chimpanzees. *Homo sapiens* (in other words, our direct ancestors, anatomically identical to us) appeared only 200,000 years ago, or perhaps earlier, but developed a mind that was the same as our own about 100,000 years later, following the evolution of new brain structures.

A few tens of thousands of years later, *sapiens* had already conquered most of the dry land on the planet. In conjunction with this event, every other human species became extinct. The same thing happened to most of the large animals, the "megafauna", which had populated the Earth for entire geological epochs and were still alive during the last glacial period. Many other animals are becoming extinct before our eyes. Indeed, our presence seems to accelerate dramatically the reduction of biodiversity, as confirmed by a recent ecological simulation (carried out at the University of Aarhus), which shows how rich and diverse the mammalian presence would have been on the planet if we, *Homo sapiens*, never existed.

As some may find it difficult to come to terms with such a long period of time, to gain at least some idea of the timescale of the phenomena we have talked about, we could scale down the universe's age of 13.8 billion years to the 6 days described in Genesis for the creation of the world. If we imagined that the Big Bang took place at the beginning of a hypothetical Monday, the first day of the Christian week, the Earth would have formed on the following Friday at about one o'clock in the morning, life would have appeared on the same day at about five in the morning and the human lineage would have broken off from that of the chimpanzee in the

2.1 The Timescales of Evolution   11

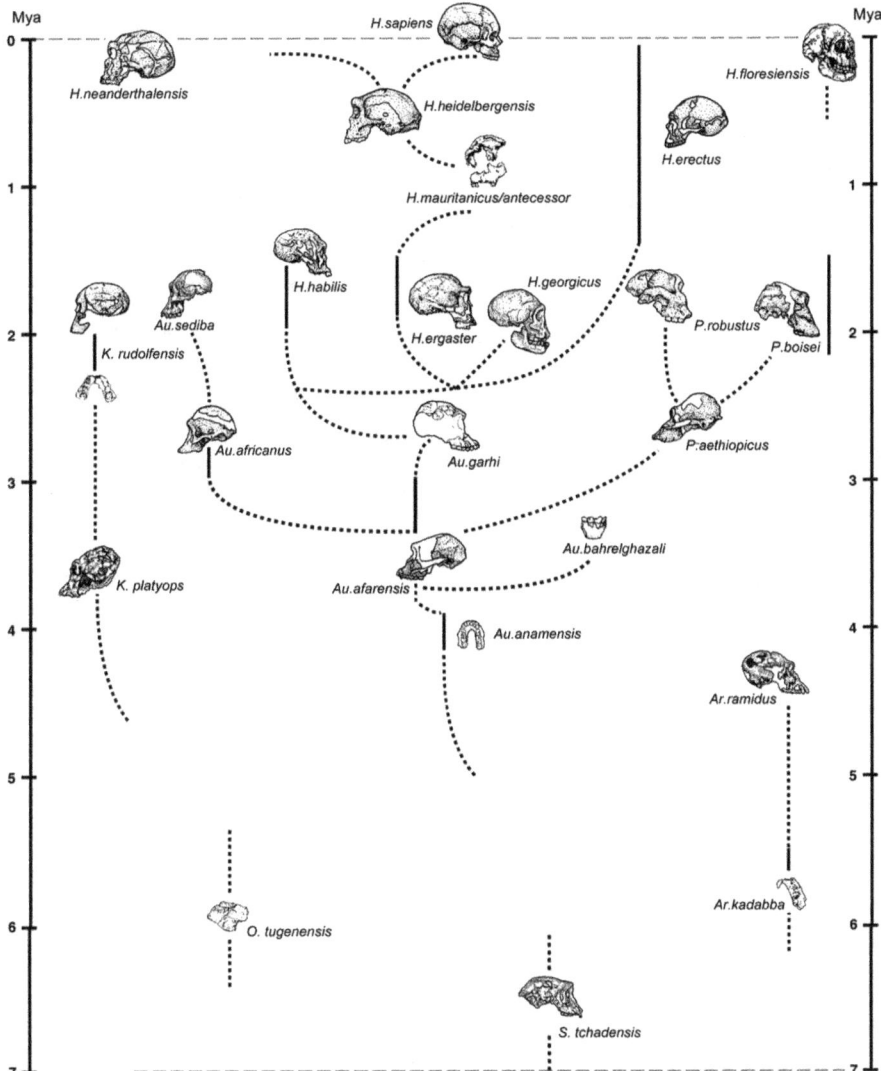

Fig. 2.1 The extended human family (suggested relationships) (MA: millions of years). *Source*: by kind permission of Ian Tattersall

final four minutes of today, Saturday night. We *sapiens* would have been born seven seconds before midnight, just a few moments ago.

From this viewpoint, paleoanthropologists joined forces with life scientists, Earth scientists and academics from other branches of knowledge to resolve the enigma of the last four minutes of our history following the trail blazed by Darwin. We have clearly not descended from the apes, at least not from the present-day apes, but we share an interesting relationship with them, as we do with all other living beings, be they animals or plants. All you have to do is go back a couple of "days" in time.

Table 2.1 Timescale of our origins (*)

	Established origin (years ago)	Chronologies scaled down to 6 days
Universe	13.8 billion	Monday, 00:00
Earth	4.5 billion	Friday, 01:03
Microorganisms	4.1 billion	Friday, 05:13
Fish, reptiles, insects	300 million	Saturday, 20:52
Mammals	200 million	Saturday, 21:55
Primates	65 million	Saturday, 23:19
Hominins	7 million	Saturday, 23:56
Homo	2 million	Saturday, 23:59
H. sapiens	200,000	Saturday, 7 s to midnight
H. sapiens in Australia	60,000	Saturday, 2 s to midnight
H. sapiens in Europe	45,000	Saturday, 1.7 s to midnight
H. sapiens in America	16,000	Saturday, 0.6 s to midnight

(*) The actual times are adjusted to fit the 6 days described in Genesis, assuming the universe formed at the beginning of the Christian week (Monday) and that it is now midnight on Saturday, Sunday being a day of rest

Chapter 3
The *Star Wars* Cantina

Science fiction often feeds upon intergalactic encounters with other intelligent life forms. Many readers will remember the Star Wars cantina, where Luke Skywalker comes across various representatives of peoples who inhabit his corner of the universe, some humanoid in appearance, others more reptilian or in the shape of improbable creatures. The purpose in the film was to play on our emotions: fear, revulsion, tenderness, amusement, but science tells us that we have actually already had such encounters with other intelligent species here on Earth.

As they wandered the planet, our *sapiens* ancestors must have encountered, at least occasionally, the Hulks who populated Eurasia, the Siberian Yetis, the Indonesian Hobbits and who knows how many other creatures of our imagination. If we could look back into our past in light of the latest scientific discoveries, it would be like watching a fantasy movie. In reality, we are talking about our true history as it unfolded in deep time.

Not long ago, we believed we were the only intelligent beings on the planet; now we know that at least three other human species, and maybe more, lived in our time. We name these species after the places in which their first remains were found: the Neanderthals in Germany, the Denisovans in Siberia, and the small *floresiensis* from the Flores Island in Indonesia. Recently, we discovered that our *sapiens* ancestors not only encountered these species but, in at least two cases, had "very close" encounters with them.

When we analyse our own DNA, that of present-day *sapiens*, we make an interesting discovery: irrespective of the colour of our skin, we are clearly of African origin, as we can all be traced back to a small group that evolved on that continent from earlier human forms.

This surprising finding is the result of analysing DNA in the mitochondria, the minute structures outside cell nuclei that provide them with energy. The mitochondrial genome, featuring a circular double-stranded helical structure, is made up of 16,569 base pairs grouped into 37 genes. These bases are given the four letters of the genetic alphabet A, G, C and T (which correspond, respectively, to the molecules adenine, guanine, cytosine and thymidine). The cell nucleus genome, on the other hand, is made up of 3.2 billion base pairs grouped into some 20,000 genes, which contain information on our anatomy, body functions and even our behaviour. It can be used to explore, for example, the difference between human species, the effect of their interbreeding, and many details of their migrations.

Fig. 3.1 Family portrait. Human biodiversity in deep time. Source: By kind permission of P. Plailly/E. Daynés-Reconstitutions Atelier Daynes, Paris

3.1 Children of Eve

Mitochondrial DNA is inherited exclusively from the mother. It can therefore be used to trace back an uninterrupted chain of genetic relationships, ideally right back to our "African Eve". In actual fact, geneticists believe that we can be traced back to an ancestral population comprising a few thousand women, counting only those who contributed to our present-day genetic structure. Each branch of the genealogical tree is identified through the use of "genetic markers", in other words, specific DNA sequences that uniquely characterise a population group (known as a haplogroup). As time goes by, random genetic mutations (changes in the letters of the genetic alphabet) build up in the mitochondrial DNA, generation after generation. These mutations can be used to identify the different lines of descent.

We can use the rate of these mutations to evaluate the time that has elapsed since two human groups with a common ancestor separated from one another. An analysis of the mitochondrial DNA of all peoples currently living outside Africa shows a very low variability. We can thus deduce that they descend from a small group of individuals, perhaps a few thousand, who lived less than 100,000 years ago in Africa. It has been suggested that a population bottleneck was brought about by the global disaster triggered by the eruption of the Toba volcano in Indonesia

74,000 years ago. This was the most catastrophic event recorded during the history of humankind: on that occasion, 2700 km^3 of magma and ash were expelled, creating a screen that shielded the planet from the sun's radiation and causing long-term climate changes on a global scale. Analyses of Antarctic ice cores suggest that this cooling period lasted several centuries.

Visiting the archaeological site of Jwalpuram in Andhra Pradesh in India, one can see layers of ash nearly 3 m deep produced by the Toba eruption. This material constitutes a flourishing business for the locals, who sell it in the markets as a metalware polish. Archaeologists are now studying the stone tools found in Jwalpuram in the geological layers below and above the ash from the Toba eruption.

Curiously enough, in 2007, in a small village in that area, one of the authors (Claudio Tuniz) bumped into an unusual character who was travelling the world on foot. He had left Slovenia, his country of origin, in 1984. He had already walked through Europe, Africa and China. At that time, he was heading for Australia and then the Americas. He was hoping to get back home by 2014. He said he had planned his 30-year odyssey to promote human brotherhood. He did not realise that he was following in footsteps that marked the entire history of our species as it spread out across thousands of generations, albeit by a slightly different route.

3.2 Human Dispersions

To get back to our story: Who left behind the stone tools found in Jwalpuram? If we cross-match the different archaeological and genetic data, they suggest that the first group of *sapiens* originating in Africa had already arrived in India before the cataclysm of 74,000 years ago. The survivors then continued in a south-easterly direction towards Australia. Recent genetic studies on present-day Australian Aborigines reveal that there were actually two great human waves of *sapiens* who crossed all of Asia from west to east. The first began their journey between 70,000 and 80,000 years ago and arrived between 50,000 and 60,000 years ago on the supercontinent, known as Sahul, which then included the present-day Australian mainland, Tasmania and Papua New Guinea. This first wave was probably slowed down, but not halted, by the difficulties of long sea crossings, but in the end, some *sapiens* managed to disembark on the Australian continent.

During the second great migration, *H. sapiens* arrived in Asia around 30,000 years ago and first populated present-day China and then all of the Far East, continuing northwards and toward the Americas through the present-day Bering Strait, which, at that time, was an emerging land known as Beringia, connecting Siberia to Alaska. Genetic analysis of Native Americans confirms that approximately 20,000 years ago, members of these two different migratory waves across Asia actually interbred before their descendants could reach the Americas.

The opportunities opened up by the ability to carry out low-cost DNA sequencing on our own genetic material, as well as that of fossil human remains, provides us with many details about when and where our migrations took place. It was discovered that no one can claim to have originated in a given area. Anyone who stopped was always bound to meet with transient populations, giving rise to descendants, some of whom stayed where they were and some of whom continued to wander the planet.

The genetic analysis of fossil remains also allows us to solve mysteries. For example, it explains why certain Native Americans have specific DNA sequences similar to those of Europeans but different from those of many Asians, from whom other Native Americans descend. It should be noted that this is not a recent inheritance. The fossil remains of two individuals found near Lake Baikal in Siberia—one who lived 24,000 years ago and the other who lived 17,000 years ago—had genes similar to those of both modern Europeans and Native Americans. Yet, they were different from those of modern Siberians and other Asians! These findings suggest that the population that inhabited northern Eurasia during the last ice age spread to the east and west, to be subsequently replaced by other Asians from the south.

The idea that humans can be considered natives of this or that region, and that they have been able to develop for a long time in isolation from one another, does not make any sense. In the long run, isolation is only a transient phenomenon.

In actual fact, we *sapiens* (all of African origin) dispersed across hundreds of generations, changing the colour of our skin and facial features in accordance with the environmental conditions and genetic drift (in other words, genetic change due to random factors). Hairy skin is normally pale. If the hair becomes sparser, for reasons we will discuss later, and solar radiation is high, natural selection will benefit descendants with darker skin, as they are more able to protect themselves against UV light. Conversely, paler skin is selected for latitudes with lower radiation, where darker skin would shut out some of the UV rays and thus slow down the metabolism of vitamin D, which is essential for the formation of bone and muscle tissue. This process does not need to be fast. Apparently, it took our direct ancestors a very long time after setting foot in Europe, 45,000 years ago, to turn from dark to pale. The gene mutation that reduces skin pigmentation occurred between 19,000 and 11,000 years ago, according to recent studies.

3.3 "Open Family" Experiments

Now that we have established that we *sapiens* met with each other many times in the course of our planetary peregrinations, what can we say about our encounters with other human species?

We know for certain that we came across the Neanderthals, a species who lived in Europe and western Asia before us. We have been aware of their existence since the mid-nineteenth century, but only now do we have evidence that we actually interbred with them: evidence that many of us carry without being aware of it. What can we say about those species that we have discovered only recently: the

Denisovans and the humans from Flores? Did we meet them as well? We certainly encountered the Denisovans and most probably the so-called Hobbits. The story is fascinating. We will start with the Neanderthals.

Because they were not very keen on harsh climates, our *sapiens* ancestors only arrived in Europe during a relatively mild period of the last ice age. By the time they arrived, the Neanderthals had already been occupying an area that extended from the Mediterranean to Siberia for 300,000 years, maybe more. They were our "cousins", because we have a common ancestor, *Homo heidelbergensis*. Even though we both descended from a remote progenitor, also of African origin, we were darker and more agile, whereas they were sturdier and paler, due to having evolved in colder latitudes.

The Neanderthals have long been represented as the primitive ancestors of us white, "civilised" Europeans. Nothing could be further from the truth. They were not our ancestors, they were not particularly primitive, and we were not yet pale. But we survived and they died out. Why was this the case? We do not know with certainty, but we have accumulated numerous clues.

Sturdy, with a stocky body, sloping forehead, protruding supraorbital ridges, and a flattened skull extended at the back to form an occipital bun or "chignon", the Neanderthal was not at all our direct ancestor, as we have already stated. We know that our direct ancestors were Africans: agile and slender, with a round skull, vertical neck and face, high forehead and pronounced chin. Just like us, in whatever bodily variant we care to assume. Our typical anatomical features probably evolved in Africa two or three hundred thousand years ago. Some paleoneurologists believe that our round skull made all the difference. It corresponds to an expansion of the brain not only sideways (this had already occurred more than a million years earlier in other human species), but also upwards due to the thickening of specific parietal areas of the brain, such as the precuneus and the intraparietal sulcus. This apparently allowed the development of better visual and spatial coordination that, when combined with our mnemonic abilities, triggered a virtuous and cumulative interaction between brain, culture and environment, giving rise to new cognitive capacities.

Even with their flattened skull, the Neanderthals were not, however, "troglodytes", despite having been portrayed as such for many years. At the beginning of the twentieth century, magazines such as "The Illustrated London News" offered the image of a fearsome club-wielding hominid who ground his teeth as he lay in wait for his next victim at the entrance to his cave.

A 60,000-year-old geological layer in a cave in Divje Babe in Slovenia yielded a bear femur together with stone tools classified as Mousterian (a name describing the Neanderthal stone industry), which had four holes spaced at precise distances. Analyses carried out using x-ray microtomography suggest that this could have been a musical wind instrument made by the Neanderthals.

This analysis technique is similar to the x-ray CT scan carried out in hospitals, but is much more powerful because it can reveal details that are hundreds of times smaller (in the order of thousandths of a millimetre). The sample revolves before a beam of x-rays, providing thousands of digital radiographs. Special mathematical algorithms reconstruct the internal structure of the specimen in three dimensions, using thousands of megabytes: nowadays, the memory capacities of computers are much larger and this is no longer a difficult undertaking. These images provide many details of the way an artefact has been worked that are not visible to the naked eye.

Although the results are not definitive, the holes in the bone do not appear to have been caused by the bite of a carnivore, as some scholars had claimed. Hyenas were assiduous cave-goers and many animal bones bear their bite-marks. We cannot therefore be sure that this perforated bone is a musical instrument, but it is nevertheless exhibited in the Ljubljana Museum as a "Neanderthal flute".

In the Fumane cave near Verona, it was discovered that Neanderthals decorated themselves with feathers removed from the birds' bodies with very sharp stone tools. Because the bones were taken from species with particularly tough flesh, such as the black vulture, red-footed falcon and other large birds of prey, it was concluded that those who hunted them were mostly interested in their fine coloured plumage. Eagle talons adapted for use as personal ornaments were recently found in Neanderthal caves in the now Italian region of Friuli and in Croatia. The talons still show the marks of the stone tools used to remove them from the birds' feet.

Cave paintings dating back to 41,000 years ago found in the cave of El Castillo in northern Spain have also been attributed to Neanderthals. Elaborate engravings with abstract motifs have recently been identified in the Gorham cave in Gibraltar and dated to 39,000 years ago. Their Neanderthal provenance seems to be confirmed by the presence of Mousterian stone tools in the corresponding cave sediments. We also know that the Neanderthals painted their bodies, probably with dark paint. We *sapiens*, on the other hand, preferred to decorate our dark skin with light colours.

A French site attributed to the Neanderthals has yielded bone tools known as *lissoirs*, used to waterproof animal skins, an invention that we *sapiens* appear to have copied from them. During these operations, it seems reasonable to suppose that they gripped the skins between their teeth, as deduced from the extreme wear on the incisors of many Neanderthals. In particular, remains discovered at La Ferrassie in France dating back 50,000 years showed total erosion of the enamel on the front teeth, to the extent of exposing the dentine and the pulp chamber. According to some, this habit could be attributed to under-development of the parietal lobe components: those governing hand-eye coordination during the execution of complex work. The teeth would therefore have acted as a kind of "third hand", used systematically to help them perform certain manual operations. In particular, a recent study on about 100 teeth (incisors and canines) from European Neanderthals showed a systematic difference in dental wear between men and women. This would indicate a division of labour according to gender.

The Neanderthals may also have had a complex language. This is consistent with a study carried out at the Trieste synchrotron on a Neanderthal hyoid found in a cave in Kebara, Israel, dating back to 60,000 years ago. This is the only bone present in the vocal tract and therefore the only element available in the fossil record. It is apparent that the internal microstructure of the Neanderthal hyoid bone from Kebara is similar to that of *sapiens*. The histological details are typical of a bone that had been subject to intensive and continuous metabolic activity. In particular, comparisons between hyoid bones of the Neanderthals and those of *sapiens* based on finite element analysis—applying the numerical procedure that engineers routinely use for detailed studies on material performance—show significant analogies between the bones' biomechanical performances. This allows us to assume that the Neanderthal hyoid bone was regularly used to utter the same sort of sounds made by our own. It naturally remains to be seen whether this combination of sounds corresponded to reasoned, intelligent speech, but the same also applies to some contemporary humans.

Recent studies have shown that the Neanderthals buried their dead and may have believed in life after death. Many anthropologists suggest that there is therefore sufficient evidence to infer that the Neanderthals were capable of some symbolic thought.

We can now use tools and methods of forensic science to improve the representation of our "cousins" and produce extremely realistic reconstructions of men, women and children. This process has revealed that they resembled us more closely than we thought, despite retaining their receding chin, sloping forehead and more pronounced nape.

What do we know of our relationship with these extinct humans? In 2010, we were finally able to sequence their genome using the remains of a 38,000-year-old femur from the Vindija cave in Croatia. In these ancient fossil bones, the DNA had been broken down into portions made up of less than 200 base pairs: the repair mechanisms normally active in living organisms were no longer in operation. Modern genetic sequencing techniques nevertheless made it possible to reconstruct the DNA's original structure.

The results showed that a small part of the Neanderthal genome, between 1 and 4%, is present in the genome of all the *sapiens* who left Africa. We must have become very close at some time! When did these close encounters take place? Certainly following our exodus from Africa, which began 70,000–80,000 years ago, but probably even earlier, when *sapiens* individuals encountered Neanderthals in the area that forms part of the current state of Israel.

The sporadic presence of both species in this region is confirmed by fossil remains found in the Qafzeh and Skhul caves. During the last interglacial period, between 130,000 and 110,000 years ago, *sapiens* spread out from south to north along the entire African continent. The Sahara was a much more pleasant environment at the time, with lakes, watercourses and abundant vegetation, as confirmed by recent isotopic analyses on wind-borne plant remains that slowly accumulate in Atlantic Ocean sediments.

Toward the end of that period, the advent of the last ice age forced various Neanderthal groups to abandon Europe to seek refuge in the milder lands of the Middle East. The possible evidence of these close encounters has been recently found: the genome of a 50,000-year-old Neanderthal woman discovered on the Altai mountains in Siberia reveals her species had interbred with *sapiens* approximately 100,000 years ago.

Our contiguity with the Neanderthals in the Middle East probably lasted longer, on and off, than previously thought, up until 55,000 years ago. This is suggested by the discovery of a *sapiens* skull of this age in the area, which was still inhabited by Neanderthals. Indeed, inter-breeding probably went on not only in the Middle East but also in Europe, right up to the very end. A recent DNA analysis of the remains of a 40,000-year-old *sapiens* discovered in Romania reveals that he probably had a Neanderthal great-great-grandfather.

Tests carried out on the hair of an Australian Aborigine, collected by a British ethnologist in 1920, show that his genome also contained parts of Neanderthal origin. Traces of the Denisovan genome were found as well. We shall return to this later.

3.4 More "Open Family" Experiments and Their Consequences

Present day sub-Saharan Africans are the only relatively pure *sapiens*, because their ancestors, who remained on the continent, did not have the opportunity to meet the Neanderthals. It nevertheless seems that their genome contains very tiny traces of another mysterious hominin, which has not yet been identified.

Now, we will take a look at the other two human species that overlapped with us.

In 2012, researchers from the Max Planck Institute of Leipzig published the genome of a new human species that has not yet been classified. Remains of this individual were found in the Denisova cave on the Altai Mountains of Siberia. The only fossil remains were the phalanx of a finger and two teeth, all dating back to around 50,000 years ago.

This human species probably inhabited a vast region stretching from Siberia to Oceania. We also know that individuals from this species interbred both with the Neanderthals and with our *sapiens* ancestors. Traces of these encounters (4–6 % of Denisovan DNA) are still present in the DNA of some of us, particularly those who settled in Tibet and present day Oceania. When, and where, did such encounters take place? The answer is circumstantial.

It is a common belief that *sapiens* and Denisovans met on the Eurasian continent between 80,000 and 60,000 years ago, during the exodus of *sapiens* eastwards. A recent discovery opens up new hypotheses for the location of such encounters; it also challenges the idea that we *sapiens* were the first to cross the Wallace line. This is a marine boundary that has always separated the Asian fauna and flora from that of Sahul, even during the ice ages, when the sea level was much lower. Hundreds of lithic tools have been found on Sulawesi, an island of present day Indonesia, in the region to the east of the Wallace line. They have been dated back to over 100,000 years. As, so far, no human remains of that age have been discovered in the vicinity, their makers remain unknown. Indeed, the lithic tools could be attributed either to *Homo floresiensis* (living in Flores since 190,000 years ago) or to *Homo erectus* (present in Java since over 1.5 million years ago) or to the Denisovans (if we rely upon the genetic information provided by the indigenous people of Papua New Guinea and Australia). If the Denisovans did populate Sulawesi, according to these findings, they would have crossed the Wallace line long before we did, and the encounters of our two species might well have taken place at the end of our journey towards Sahul.

In any case, the hybridisation among species is not only a matter of curiosity. It bears important implications, both negative and positive, for our current health. We will talk about the negative effects of our interbreeding with the Neanderthals in Chap. 9. Here, we will briefly mention some positive effects. Certain Neanderthal genes helped us to adapt to the cold and to the low levels of UV light in northern Eurasia. A variant of a gene, referred to as EPAS1, inherited from the Denisovans, turned out to be useful to the *sapiens* who settled in Tibet. By saving on the amounts of oxygen needed, it helped them adapt to altitudes over 4000 m. The ancestors of the future inhabitants of Oceania had also genetically encountered the Denisovans. But as EPAS1 served no purpose on the Australian plains, it was probably lost through the effect of natural selection.

Finally, it seems that another human species crossed the Wallace line before us. In 2003, the remains of some individuals of a fourth, bizarre human species were found in a wonderful cave on the island of Flores, set between Bali and Timor, amidst natural surroundings seemingly borrowed from the set of a fantasy film. These were classified as *Homo floresiensis* and immediately nicknamed Hobbits, after the heroes of Tolkien's books.

In the beginning, it seemed impossible that these people could have been human. They were no taller than one metre in height and weighed no more than 25 kg. They had enormous flat feet, long arms and a brain that was about the same size as that of a chimpanzee. However, they manufactured stone tools to hunt large animals. It is known that they fed on giant rats; they also hunted the *stegodon*, a species of dwarf elephant-like creature that is now extinct, and even the large monitor lizards measuring 6–7 m in length that still survive on the northern coast of Flores and on the offshore islands of Rinca and Komodo. The latter species is extremely dangerous due to its poisonous bites, which cause a slow and extremely painful death. It is difficult to imagine how, and with what courage, this species of tiny hominins was able to hunt and eat such animals.

When did this human species disappear from the face of the Earth? Initial radiocarbon and luminescence dates suggest that they survived up to 17,000 years ago, but this chronology has been revised in 2016. New dates based on different geo-chronometers, including uranium-thorium, show that their bones and stone tools disappeared from the fossil record around 60,000 and 50,000 years ago, respectively. The demise of these humans coincides with the arrival of *Homo sapiens* in South East Asia, but our role in their extinction is not clear. "It's a smoking gun for modern human interaction, but we haven't yet found the bullet", said Bert Roberts, one of the scientists involved in this study.

In the beginning, when people were reluctant to believe that an actual new human species had been found, some suggested that they were the remains of some deformed *sapiens* who had been struck by a strange disease, such as microcephaly, a neurological malformation that causes the brain to remain very small and stunts growth.

This reaction is quite typical and crops up regularly whenever a new hominin species is discovered. It is understandable that some members of the academic community would wish to defend their scientific views vigorously if the new discoveries shed doubt on it. We need only think back to the discovery of the first Neanderthal remains in 1856 when some medical doctors immediately suggested they belonged to a *sapiens* with rickets.

The doubts shed over *H. floresiensis* seem to have no substance. Yet, it was recently proposed that the remains found in the Liang Bua cave were those of a small *sapiens* with Down's syndrome, a disease that makes the brain somewhat smaller, shortens the limbs and flattens the feet. The scientific community kept rejecting such claims. This was, indeed, the third time that a degenerative disease had been suggested as an explanation for the existence of these new hominins. A famous Australian anthropologist said at the time: "The discoverers seem to lack the capacity to recognise a village idiot when they see one". But as more and more hominins of this kind were subsequently discovered, it would suggest that there was an entire village of idiots!

Other paleoanthropologists believed that these remains belonged to a new species of archaic human, similar to *Homo erectus*, who had dwarfed due to living on an island where resources were scarcer. The closest analogies are those of dwarf elephants in Sicily and dwarf mammoths in Sardinia in the ice ages. Some scholars went on to suggest that they were a variant of *Australopithecus*. It has unfortunately not yet been possible to analyse their DNA, despite recent attempts by two research groups. Who knows whether a Hobbit fraction could turn up in some of us? It seems unlikely, but on the other hand, our behaviour never ceases to amaze. We only need to think of events that some believe may have occurred concerning our relationships with the apes in deep time.

Chapter 4
The Apes and Us

Some anthropologists think that interbreeding took place, perhaps for relatively long periods, between some of our early ancestors and those of the chimpanzee. But for this line of conjecture, the arguments are very questionable and disputed. These events can apparently be traced back to the period following the separation of the two lines of evolution from the common ancestor. The suggestion of a certain reciprocal "affection" between our ancestors and those of the chimpanzees, which would then have been morphologically and genetically closer to one another, is based on a discrepancy between the dates on which the final separation between the two evolutionary lines must have taken place.

The dates obtained from genetic studies differed from those emerging from archaeological studies: according to genetic data, the separation took place about 6 million years ago, but the age of the earliest hominin fossil remains, belonging to *Sahelanthropus tchadensis,* date back to 7 million years ago. Some suggest that this discrepancy could easily be explained by assuming that genetic interbreeding may have carried on after the separation. This would have reduced the apparent genetic divergence between the chimps and us. This divergence is based on the number of DNA letters that are different between the two species; it can be converted into "divergence time", by assuming that the mutations that create this difference between genomes occur at a constant rate. Unfortunately, the accuracy of the molecular clock is questionable and the archaeological dates are still tentative. Some even doubt that *Sahelanthropus* was a real hominin. As a result, we cannot reach any definite conclusion according to this line of reasoning.

There is a new twist to this story. It was recently discovered that human and chimpanzee X chromosomes are markedly less different than are the other chromosomes. This fact induced some geneticists to suggest that, about 4 million years ago, male proto-chimps mated with female proto-humans. The male hybrids would have been infertile (as can be explained by known genetic rules), but the females would have been fertile and could have backcrossed with male proto-humans, and thence onward.

4.1 The First Scandals

The feeling of a certain similarity between us and the other great apes is, indeed, difficult to ignore. When Darwin saw an ape for the first time, he noticed that she behaved "like a naughty child". Jenny was the first young orangutan bought by the London Zoo, and had been dressed for the occasion in a flowered pinafore. These animals provoked both fascination and revulsion in conformists of the age. Queen Victoria, for example, found them "disagreeably human".

Society at that time, particularly the part closest to the church, would not have been kind to Darwin. We have already spoken about how he was ridiculed in the press by caricatures showing his bearded face on the body of an ape. Yet, it would have taken more than slander and invective to discourage all the naturalists and adventurers who immediately began to seek the "missing link" between us and the apes, a misleading concept that still lingers on in the minds of many to this day. One such believer was the Dutch physician Eugène Dubois who discovered Java Man in 1894, naming him *Pithecanthropus erectus* in the conviction that he was an "ape-man", although he was eventually reclassified by others as *Homo erectus*. This individual dated back to approximately 500,000 years ago.

The academic world of the time displayed the usual scepticism that accompanies every new discovery, suggesting that the find was that of an extinct gibbon. Dubois nevertheless remained convinced that it was the elusive "missing link" and that the cradle of humanity was in Asia. On the first point, he was (partly) right: it was indeed a hominin, albeit not the missing link. His second assertion nevertheless proved to be completely wrong.

Thirty years later, Raymond Dart found the Taung child in South Africa. This was an even more ancient hominin dating back, as we now believe, to approximately 3 million years ago. Anthropologists of the period did not initially agree that this find belonged to our evolutionary line either; instead, they considered it to be some kind of ancient extinct ape. It nevertheless turned out to be a proper hominin, and the species it represents is now known as *Australopithecus africanus*. Only later was it found that *Homo erectus* was of African origin, when similar and even more ancient humans, known as *Homo ergaster*, were found on that continent. Java Man was actually their direct descendant, the result of their previous migration eastward.

Our connection with the apes is the aspect that has attracted most headlines and led to the most enmities towards the theory of evolution. Often, the most outrageous reactions stem from misleading interpretations and preconceptions that persist to this day. Indeed, we now know that we share only one ancestor with the chimps. In any case, he was neither a man nor a chimp. So when was the period of greatest expansion of the now-extinct great apes that we also call hominoids? This is not a digression. A very brief overview of this evolutionary trajectory will actually help us to better understand ourselves, the most recent of hominins.

4.2 Planet of the Apes

To answer the question, we must refer to changes in environmental and geological conditions during the deep past. How can we reconstruct these conditions for such a long and remote period?

The history of how environmental conditions have changed on Earth is actually stored in many natural archives. A record stretching from the earliest times up to the present day has been preserved, for example, in the rocks (as far back as the origins of the planet), in marine sediments (up to tens of millions of years ago), in the long cylinders (known as cores) of ice that can be obtained by drilling the polar ice caps (up to 1 million years ago) and lastly, for periods closer to our own time, in lake sediments, stalactites, corals and even tree rings. In the latter case, we can only go back to the last 10,000 years.

Many details can be read in all these natural archives. For example, different oxygen isotopes are preserved in every layer of marine sediments, Antarctic ice cores, corals and stalactites. All you have to do in order to obtain the temperature during the corresponding period is calculate the ratio between concentrations of two oxygen isotopes, oxygen-18 and oxygen-16. In lake sediments, you can also identify pollen grains and hence plants from the period corresponding to that layer.

The history of the apes, on the other hand, is written in their fossilised bones and teeth. Once, these findings were few and far between and analytical techniques were based only on observation and morphological comparison. Nowadays, we are able to find many new remains, sometimes through the use of advanced satellite remote sensing. We are also able to analyse their internal microstructure using quantitative and non-destructive techniques. When we cross-match all this information, it recounts the story of the "Planet of the Apes": what the Earth became between approximately 25 and 15 million years ago.

Fig. 4.1 Chimpanzees talk it over in committee. Shutterstock.com, Copyright Patrick Rolands

At the time, Africa and Eurasia had become linked due to the movement of their respective tectonic plates. The high global temperature caused immense evergreen forests to grow up and ultimately cover most of the dry land. More than 100 species of hominoids were then dispersed between the Iberian Peninsula, China and southern Africa. But things were about to change.

Long before, about 50 million years ago, the Indian plate had crashed into the rest of Asia, forming the Himalayas and raising the Tibetan plateau. This brought about a change in atmospheric circulation and favoured the absorption of carbon dioxide by the newly-formed rocks. From that time, the overall temperature of the Earth had begun to drop. But this very long-term trend was broken and reversed at the time we are interested in, ie, between approximately 25 and 15 million years ago. After this warm period, the global temperature began to fall again. Gradually, many areas became arid, splitting up the apes' habitats. This put an end to the Planet of the Apes, since most of them subsequently became extinct, although some adapted to the new environments.

Very little remains today of that biodiversity. In addition to the gibbons, only two species of orangutan survive in Asia. These are found in Borneo and Sumatra. Two species of gorilla, in addition to chimpanzees (common chimpanzee and bonobo), are clinging on in Africa, but we do not know for how long. Yet, the last surviving species of *Homo* has colonized the entire planet. When did the Human Planet begin to form, and how?

4.3 The Human Planet

The geological record tells us that 6–7 million years ago, when the Planet of the Apes was almost over and the forests began to give way to savannahs in extended areas of Africa, some bipedal creatures made their first appearance. They were different from us and also from the other surviving great apes. These are often called "proto-humans" or "pre-humans", because they evolved before the genus *Homo*. It has been suggested that the upright posture began randomly, within a particular group of hominoids, and was then imitated and transmitted to subsequent generations because it was more suitable for survival. Only later would natural selection favour the individuals best able to absorb this innovation to the point of transforming it into a genetic and anatomical adaptation. Though we know little about our ancestors, the mists have begun to clear in recent years.

One of the forms that preceded humans included Ardi, a female living in a period very close to that of the common ancestor that we shared with the chimpanzees. Her brain measured 350 cubic centimetres; she weighed 32 kg and stood 120 cm in height. The bones of this *Ardipithecus ramidus*, as she was classified, were discovered in the Afar Depression, in present-day Ethiopia, one of the lowest areas in Africa, sandwiched between layers of volcanic deposits. Because ash always contains a small percentage of potassium, including one of its radioactive forms (potassium-40), the age can be determined with certainty because potassium-40 decays very slowly to turn into an isotope of argon, argon-40. By counting the argon-40 atoms present in the volcanic material using a mass spectrometer, we can therefore determine Ardi's age with great accuracy: 4.4 million years.

Although her new iliac structure allowed her to walk upright, she still retained longer arms and relatively prehensile feet with an opposable big toe. This meant she could still climb trees.

Going back in time to the common ancestor between us and the chimpanzees, hominins became increasingly different from present-day humans but no more similar to chimpanzees. Their evolutionary adaptations proceeded independently and sometimes they developed particular traits that did not concern us at all. For example, *Ardipithecus* did not have the modern specialisations developed by the chimpanzee, such as walking supported by the knuckles of their upper limbs.

Lucy, a female *Australopithecus afarensis*, also bipedal, appeared approximately one million years later. This species is now known by abundant material from Ethiopia, and some from Kenya too. Lucy was more specialised than Ardi and more inclined to walk upright, even though she must still have been a climber. She had longer legs, a big toe that was no longer opposable and a foot that was rigid but already slightly arched, almost like our own. However, she probably still walked slowly and unsteadily. At that time, numerous hominins of different species were in existence and lived a few kilometres away from one another in the Afar Depression. Among these, the recently discovered *Australopithecus deyiremeda* already had structural dental and jaw characteristics that are traditionally associated with the genera *Homo* and *Paranthropus*.

We must nevertheless not fall into the trap of believing that Ardi, Lucy and the other hominins who came before us were unfinished beings, freaks of nature or prototypes that had to be improved upon to reach our present form. Their bodies were actually perfectly adapted to their natural habitat: nowadays, we would call them "successful solutions". They, like us, were the combined outcome of chance and necessity. They inhabited this planet for millions of years. So far, we *sapiens* have only managed some 200,000 years, even less, a mere 100,000, if we go back to when we acquired the mental attributes that characterise us today. As the genus *Homo*, we have nevertheless lived to the respectable age of more than 3 million years, according to recent discoveries.

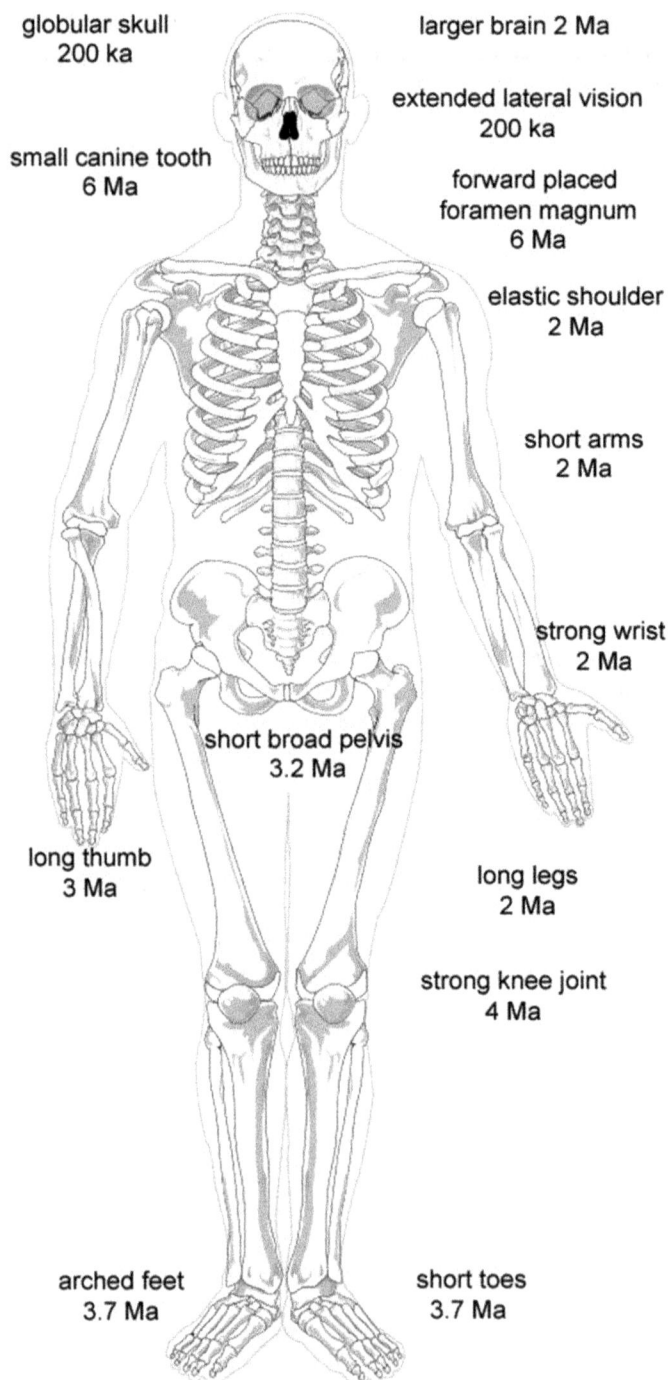

Fig. 4.2 Timeframe of anatomical changes during human evolution (Ma: millions of years; ka: thousands of years). *Source*: Adapted from Wikimedia Commons

Chapter 5
The Quest for Fire

We have already mentioned the successful solutions adopted by the various species of hominin who came before us. Sometimes these took the form of stronger teeth and jaws, while on other occasions the difference was the result of body size, indiscriminately smaller or larger, depending on the environmental circumstances. In some cases, the solutions concerned the best systems of locomotion or other specific physical characteristics (the ability to climb or run). We nevertheless appear to have ruled out the approach taken by the large predators of equipping themselves with "built-in" weapons, such as more powerful and sharper claws.

It now seems reasonable to ask: What were the successful solutions adopted by the many variants of *Homo* at the dawn of humanity? We can start to answer this question by considering new sources of advantage in the struggle for survival. It could be useful to focus on one important step that allowed us to improve our ability to survive without changing our anatomy.

5.1 From Scavengers to Opportunists and Hunters

At a certain point, we became aware that we could equip ourselves with stone objects that we were able to collect and shape using our own free hands. They could be used to scrape the last scraps of meat from animal carcasses or to break bones in order to get at the marrow, but they could also be used as weapons for killing small animals or other hominins. Initially they took the form of pebbles with a single cutting edge.

It was always thought that the most ancient tools of this type, created in an archaeological industry we call Oldowan, dated back to 2.6 million years ago. In 2015, however, similar stone tools were found in Kenya (near Lake Turkana) dating back 3.3 million years, 500,000 years before the official appearance of *Homo*, whose earliest fossil remains, dating back 2.8 million years, were also discovered in 2015. According to their discoverers, they could be attributed to pre-human hominins such as the species *Kenyanthropus platyops*, which, at the time, lived in that still-forested area of Africa. In this case, we might lose our right to lay claim to this innovation, which is considered a cognitive leap crucial to the advent of the *Homo* species, ascribing human traits to other genera of hominin. Alternatively, we might attribute these artefacts to a new, as-of-yet undiscovered *Homo* species, evolved before the onset of the glacial cycles.

The first tools with sharp edges on both sides appeared 1.8 million years ago and are known as Acheulean tools. These artefacts, which were definitely produced by the *Homo* genus, were obtained by working the stone repeatedly on both faces in order to obtain a fairly flat object. The result was a very sharp almond-shaped knife. In some cases, this involved a long and difficult procedure that only a few of us would be able to undertake nowadays without some practice. At this point, the tools were much more than mere "cutlery" used to glean a few scraps of food from the leftovers of others' meals, but could be used as weapons proper. With the addition of a wooden handle, they actually made excellent axes.

Another formidable ally that came onto the scene during the same period in which these more technologically advanced tools first appeared was fire. Control of fire was to have a decisive influence on human evolution.

Wildfires already had a significant impact on the environment frequented by humans, whether this was the dry, arid African savannah or the icy Eurasian steppes. They soon learned to use it to light the nights, keep warm and socialise, but also as a weapon for defensive as well as offensive purposes.

Many argue that moving from a diet based on plants and insects (such as that predominating in a forest) to a more omnivorous diet that included otherwise indigestible foods (such as tubers) and meat was only made possible through the control of fire. However, recent findings suggest that stone tools were used by early humans for processing meat and nutritious roots or tubers, long before the development of cooking. Fire also permitted hominins to keep large carnivores away from the prey they had just killed and to survive behind their backs. Fire also enabled them to hunt prey directly, cook it and digest it with greater ease. This led to greater brain development, an evolutionary solution that was to prove strategic to our survival. Fire also made it possible to sterilise meat against the potential effects of all the pathogenic parasites and germs that are present in raw food. During this process, we rose from a medium-low position to a higher position in the food chain. We were not yet competing with the large carnivores, but at least we were turning from prey to predators.

The Acheulean tools we mentioned above are attributed to *Homo ergaster*, the first fully human species, which appears in the African fossil record of about 2 million years ago. He had a much more voluminous brain than previous hominins. Not only could he walk upright, as the australopithecines had already been doing for millions of years, but he could also run at high speed over long distances due to his new Achilles tendons, his longer femur and other anatomical innovations. Natural selection also improved his hands, which in previous hominins had already become increasingly adapted for gripping objects and possibly throwing them.

Interestingly, a recent study shows that our hands are not too different from those of the common ancestor we share with the chimps. In any case, the *ergaster* hands already had an elongated thumb, shorter fingers, an extended gripping surface and all the necessary tendons and muscles. They had become precision instruments, designed to work stone tools and, if necessary, to equip them with a handle. With a stronger and more flexible wrist, these objects could be grasped firmly and used for every need. This made it possible to dig into the ground in search of roots and tubers, extract the marrow from the bones of prey and also strike animals and other adversaries from a distance. The latter capability was also enhanced in *ergaster* through anatomical changes to the shoulder that made it possible to store the elastic energy required for throwing projectiles.

The major difference, though, was almost certainly made by the control of fire.

5.2 Climate and Evolution

Over the last 3 million years, we have gone through several glacial cycles of different intensity and duration. Cooler phases have alternated with warmer phases at a pace dictated by astronomical cycles, which depend on changes in the Earth's orbit around the sun. The cold phases could last from 40,000 to 100,000 years; the warm phases from 20,000 to 30,000 years.

It has been noted that this period of great environmental changes coincided with a period during which natural selection favoured hominins with a larger brain, allowing them to build up relationships within increasingly large groups. We will return to the importance of this point below.

While it was freezing in Europe and Asia, in Africa, the drought led to fires. The forests dwindled in the north and south, making way for steppes and savannahs, respectively. In Africa, on the edge of these increasingly sparse forests, the latest robust australopithecines could survive (but not for too long) by living on roots, berries and other low-quality foods.

Fire became a constant presence, particularly in the Great Rift Valley, the lowland region that runs through nearly all of East Africa. All animals were afraid of fire, except for *Homo ergaster*, which eventually learned to master it. Of course, we do not know how this happened, but we can be sure that in order to do so, these humans had to overcome an innate fear that no other animal has yet been able to dominate. They had to develop the ability to control their emotions. They must have dreamed of becoming "masters" of the fire: creatures that did not then exist in nature. This allowed them to survive and evolve, overcoming many difficult periods. Thus, we see the start of the biological and behavioural changes that almost 2 million years later would lead to modern *Homo sapiens* and other human species.

How far back can we trace the first use of fire? And which species were responsible for it? The most ancient human site that allows us to answer these questions, so far, is that of the Swartkrans cave in South Africa, where archaeologists have found bones burned by non-spontaneous combustion dating back to 1.5 million years ago. These early humans, probably *H. ergaster*, must have realised that fire could be used not only to heat but also to catch and cook new prey.

Armed with fire, advanced stone tools, a larger brain and perhaps better communication skills, the *ergaster* were the first humans to leave Africa, about 2 million years ago. They gradually reached West Asia, China and Southeast Asia, where their fossil remains date back to 1.8 million years ago, in an Asian variant known as *H. erectus*. Hearths dating back some 800,000 years have been found at Zhoukoudian in China and attributed to this hominin or to a related species, *Homo pekinensis*.

The use of fire was already an advanced practice for one descendant of *ergaster*, *H. heidelbergensis*, a human with a 1200-ml brain who was subsequently to evolve into *H. sapiens* in Africa and *H. neanderthalensis* in Eurasia. Their hearths have been found in France, containing the remains of burned bones, blackened stones and even a kind of chimney. The remains of this species date back to between 350,000 and 600,000 years ago, a period spanning various glacial and interglacial cycles. At that time, Europe was roamed by the ancestors of present-day rhinoceroses and buffalos, which must have tasted excellent when roasted in the newly-invented hearths with chimneys.

The lifestyle of the Neanderthals and modern *sapiens* also required an intensive use of fire. Under its light, the *sapiens* were able to represent an incredible variety of animals on the walls of their caves. Because they were drawn using pieces of charred wood, it was relatively easy to date these paintings without damaging them. The rock art of the Chauvet cave in France dates back to more than 35,000 years. In this case, archaeologists used the radioactivity of a particular form of carbon, carbon-14, which is produced by nuclear reactions between cosmic rays and the atmosphere. It is based on the decay of carbon-14 atoms in organic materials. Modern radiocarbon dating methods achieve such a level of sensitivity that a fraction of a milligram is sufficient to perform the analysis. This nuclear chronometer only works for the last 50,000 years, but that is all we need to study the arrival of *sapiens* in Europe, Australia and the Americas.

It may happen that calcite precipitates hide the paintings when made on the ceilings of a cave. Fortunately, uranium-series dating can be applied, in such cases, revealing their formation process. In the case of the Altamira cave, in Spain, such a method shows that some of the underlying pictures were older than 35,000 years and went on being painted during a prolonged period of time, between 35,000 and 15,000 ago. It was a collective endeavour—passed on through generations—that well deserves its World Heritage listing.

Gathering around fires, *sapiens* could also produce music, using the long bones of the griffon vulture and mammoth tusks marked by holes bored into them with stone tools. Accurate radiocarbon dating has shown that some, in Hohle Fels, Germany, date back to 43,000 years ago.

5.3 Weapons of Mass Destruction

It seems that modern *sapiens* were able to do a little more than Neanderthals with fire: they could change their environment—and they wanted to. They began with the extensive and selective use of fire for hunting. This behaviour, which immediately had a devastating impact, has not yet been recorded for any other human species.

The environmental impacts of *sapiens* have been documented in studies of Lynch's Crater, an ancient volcanic lake in northern Australia whose sediments harbour the secrets of the last 200,000 years of environmental history. The various layers, also dated by means of natural radioactivity, contain pollen that describes how the Australian flora has changed over time.

Before the arrival of *sapiens*, the geological records clearly show a switch from rainforest to the first eucalyptuses, which love fire and even reproduce better if they are burnt. In this case, the change was determined by a selective adaption that took advantage of the wildfires that occurred in Australia due to the drought. Yet, the sediments of 40,000 to 50,000 years ago show a considerable increase in charcoal particles coinciding with the arrival of *sapiens* on the continent. By freeing up vast tracts of forest using fire, the newcomers could effectively coordinate their hunting techniques and control animal movements. In doing so, they caused a dramatic ecological impact on both flora and fauna. The eucalyptus tree has dominated the Australian flora ever since.

Fig. 5.1 First humans in Australia: hunting Diprotodon. *Source*: Drawing by Tullio Perentin, ZOIC, Trieste

Archaeological records extensively document the fact that large animals lived undisturbed in Australia for millions of years: *Thylacoleo carnifex*, a marsupial lion, *Diprotodon optatum*, a giant rhino-sized marsupial with a camel's nose, *Genyornis newtoni*, a wingless bird weighing more than a tonne, and *Megalania prisca*, a 6-m-long lizard. All these animals were wiped out upon the arrival of *H. sapiens*.

Similar stories may be recounted for the Americas, when the sudden appearance of modern humans, approximately 15,000 years ago, coincided with the disappearance of *Smilodon fatalis*, the sabre-toothed tiger, the mighty camel *Camelops hesternus* and other large animals of the glacial period. An ancestor of the present-day elephant, *Gomphotherium*, may have been one of the last victims. The bones of this prehistoric animal were recently found lying near arrowheads attributable to the Clovis culture, which characterised one of the first *sapiens* groups colonising the Americas.

And these were not isolated cases. Following our arrival on each island or continent that had not yet experienced our passage, we would bear witness to the systematic disappearance of species and genera that had lived undisturbed for millions of years despite numerous and dramatic climate changes. They had not learned to fear us. Within a couple of thousand years in Australia, 23 out of 24 known species weighing over 50 kg died out, along with countless other lighter species. The entire structure of the continent's food chain was altered. Some scholars blame climate change, but recent palaeoclimatic and archaeological studies confirm the human contribution to the slaughter. In particular, the discovery of a huge number of burnt eggshell fragments across Australia reveals how *Genyornis* were driven to extinction, around 47,000 years ago: it was not only through our hunting them, or destroying their habitat; it was also through our mass feeding on their eggs.

Similar extinctions occurred in North America, where 34 out of 37 large mammal genera disappeared in a short time span, around 11,000 years ago. In this case, though, climate change could have contributed to their demise. In South America, 50 out of 60 megafauna genera disappeared. The same phenomenon was to take place in Madagascar where the giant elephant bird, *Aepyornis maximus*, disappeared soon after our landing 2000 years ago. In New Zealand, the big moa bird vanished after the arrival of the first Maoris between 800 and 1000 years ago. In the Caribbean, the giant sloth disappeared following our arrival 5000 years ago. The dodo, *Raphus cucullatus*, a bird endemic to Mauritius, met the same fate upon the advent of the Portuguese in the seventeenth century. The list of "casualties" goes on for all the islands or isolated continents on which we set foot.

The extinctions of large animals in Africa proceeded at a slower pace. Having evolved with us, they learned how to deal with us. But that did not ultimately help them much. It is, indeed, estimated that most of them will disappear within this century along with our closest relatives, the last apes. The only animals of a certain size that get along with us are those that we raise as foodstuff, those that help us get around, for work or leisure, and those we keep as pets. All the others tend to be wiped out by hunting or by devastating their natural habitat.

5.4 The Disappearance of the Neanderthals: Many Clues, Little Evidence

Even the Neanderthals succumbed to our arrival in Eurasia, despite the marginal episodes of genetic interbreeding discussed previously. Recent studies have greatly narrowed down the period when we lived alongside the Neanderthals. Although the coexistence was initially believed to have lasted longer than 10,000 years, very recent analysis of hundreds of samples taken from dozens of archaeological sites, from Spain to Russia, show that we may have cohabited—on a steady basis—for less than 3000 years. This is a very short period when we consider that the Neanderthals were another very ancient and well-acclimatised human species. The territorial replacement of Neanderthals by *sapiens* did not happen suddenly, but rather proceeded gradually at different times and in different places in accordance with a mosaic pattern that steered the different Neanderthal communities further and further apart, thus exacerbating the ongoing underlying biological and cultural separation. The last remnants of their culture disappear from the archaeological record about 40,000 years ago.

According to some scholars, the Neanderthals had already undergone a profound demographic crisis before our arrival, having scaled down to no more than 70,000 individuals (with only a few thousand breeders) increasingly separating into smaller groups. Under such conditions, it is reasonable to suppose that close inter-breeding could have been commonplace, as was indeed confirmed by a DNA analysis carried out on the remains of the Neanderthal woman from the Altai Mountains. This condition could have added a genetic disadvantage to the list of their impairments. We cannot, however, rule out our responsibility in the final act of this drama. We do not know how many of us had arrived in Eurasia by that time, but both genetic analysis and archaeology suggest that our population grew dramatically in the corresponding period.

Some argue that one of the reasons for the Neanderthals' disappearance was because our presence reduced the resources they needed for survival. Yet, too few of us were around at that time to support this hypothesis: the planet was still capable of sustaining both species. It is more likely that the exercise of excessive violence that distinguishes us, and that has led to the genocide of so many peoples different from our own, was already at work. Tolerance does not seem to be one of our genetic traits. The Neanderthals were probably a species too similar to us to ignore, and too different to be tolerated.

Most likely, in order to prevail, we had to rely upon advantages that the Neanderthals did not have (eg, the capacity to form large bands, or the availability of better weapons, such as throwing spears, as opposed to those suitable for close-body combat).

The destruction of the natural environment continued when, once the Big Chill was over, we devoted ourselves to agriculture and livestock farming, reducing biodiversity to the limit. We also used fire to extract copper, iron and other metals from stone and ultimately to forge ever more powerful weapons, this time to be used against other *sapiens*. The last 1000 years of our history are full of genocides attempted or perpetrated against the most vulnerable peoples.

At this point, we had reached the pinnacle of our ability to dominate other living creatures and nature in general. Our potential for destruction seemed to have no bounds. Some say that we had achieved this position too quickly, from an evolutionary viewpoint, and that we were not yet able to manage it. Yet, once most of the large animals had disappeared, the *sapiens* who settled in Australia learned how to live in harmony with the continent's ecosystem by protecting the flora and fauna from further damage until the Europeans arrived two centuries later. This means it is not impossible to learn from our mistakes, but we need to determine the conditions under which we are able to do so.

In recent centuries, our appetite for energy led us to start burning coal and other fossil fuels, contributing to global climate change. In the past, at the brink of extinction, we managed to face dramatic climate changes, and even turn them to our advantage. The next challenge is to see whether we can survive and prosper through the environmental disasters and exterminations that we ourselves are now able to unleash.

Chapter 6
The Naked Ape

During the course of our evolution as *Homo*, we have quite clearly distinguished ourselves from the other animals not only through our ability to control fire but also through changes to our bodies that, at first glance, do not seem to make any sense at all. To begin with, the idea of walking upright is hard to justify. It certainly makes us unstable, exposes our vital and reproductive organs, narrows the birth canal, and means we are afflicted by several ailments as we get older. Even our young initially prefer to crawl on all fours before they try to walk like their parents. To explain why we put up with all these disadvantages, we have to look for compensatory benefits: for example, the ability to run faster, to consume less energy, to see further, and to cover open spaces and leave our upper limbs free. The topic is nevertheless still wide open for debate.

A second, much more intriguing mystery is why we progressively rid ourselves of our hairy pelts. After all, these helped us to survive in a wider range of temperatures, they protected us against the knocks and bumps of our adventurous lifestyles, they shielded us from the sun and they allowed us to wander around with our young holding onto our pelts so that we were free to use our upper limbs, and that is just a short list of benefits.

What countervailing benefits could possibly justify our current light hair state? The only mammals that seem to have embarked upon this path are the ones living in the sea, such as whales and dolphins. In this case, it is clear that hair is not needed. The fact that it is useful to us is proven by the fact that in colder climates we have always been eager to wear the furs of other animals and in warmer climates have almost always sported some piece of clothing, however small.

What is the explanation for losing our hair and when did we start to wear clothing? Admittedly, we have not lost our hair completely. We kept it on our heads and also where we needed to prevent excess sweating that would cool the body down: in the armpits and groin, for example. Our body is also still covered by a fine layer of hair (which is not actually so fine, in some cases).

If we ignore these details, the questions are still difficult to answer because, unfortunately, hair is not preserved in the geological record and neither is clothing.

Recent studies using electron microscopy and x-ray microtomography have allowed us to see the imprint left by the hair of some animals in the Pleistocene. What remains are empty microscopic channels left by fossilised hairs in the rock. Regrettably, more finds of this kind are unlikely to shed any further light on the conundrum.

Some have recently suggested, somewhat boldly, that the very fact of our hair loss (probably as a result of an individual genetic mutation) led us to evolve an upright posture. If you are a mother, it is undoubtedly difficult to carry offspring as you go about your day-to-day activities without having a hairy pelt to which they can cling. According to this view, the "state of necessity" determined by motherhood—the need to hold your young in your arms—was instrumental in freeing up the upper limbs from the functions of walking. Yet, this gives rise to a disadvantage, affecting only the female gender, when it comes to survival-related activities. This supposedly led to female dependency on males for their livelihood and, consequently, female sexual "gratitude" to the male, a condition that would have given rise to the first families. This conjecture is still, however, very debatable and has found little support in the scientific community.

6.1 The Loss of Hair: A Few Facts and Many Hypotheses

We'll begin with the known facts. Of all the existing great apes—gibbons, orangutans, gorillas, chimpanzees, bonobos and *H. sapiens*—we are the only ones without an evident hairy pelt. Before we try to understand why this should be, let us take a look at when this story began. Considering our earliest forefathers, the loss of hair probably did not affect *Ardipithecus* or the other bipedal apes that were still living in forest habitats 6 to 4 million years ago. The process is more likely to have begun specifically with *Australopithecus*, a genus typified by having abandoned a predominantly arboreal life to embark on the adventure of walking with a bipedal gait.

The fact that the hair loss coincided with an increase in the number of our sweat glands seems to confirm the need to adapt to a warmer, drier climate in order to develop a more efficient system for cooling the body. A life led close to water would also have made hair quite superfluous. In this case, any reduction in our hairy coats would have had to go hand-in-hand with the formation of a layer of fat designed to make us partly waterproof. And this is quite literally what happened.

One school of thought actually attributes some of our physical and behavioural characteristics to a presumed "aquatic stage". Although these suggestions have little support, they still deserve our attention because they are sustained by a variety of arguments: these include our ability to swim at an early age, the capacity to control our breathing, the presence of glandular systems capable of producing grease and tears, and the aptitude for mating face-to-face. All of these characteristics are similar to those of aquatic mammals. These arguments would be compatible with an interlude of survival in isolation, in a marshy, lake or marine environment such as that prevailing in the vicinity of the Afar Depression in East Africa about 6 million years ago. This is the very time when we split off from the chimpanzee. Only later, when we had lost all our hair and become "aquatic and bipedal", did we adapt, according to this theory, to the drier savannah conditions that are more commonly described.

In any case, all the evolutionary solutions mentioned above are compatible with the environmental changes prevailing during the peak period of the australopithecines, who, for millions of years, witnessed the advance of savannahs to the detriment of forests and were subject to an increased need to gravitate to the vital water sources. The hair loss may also be explained by reasons related to sexual selection, which would arguably make hairless individuals more attractive, particularly in the case of women.

There is also another explanation that could seem more trivial. Less hair led to fewer parasites, which could easily escalate from a mere annoyance to a torment, particularly in warmer climates. Ironically, our parasites can help us to understand the loss of our hairy coat and the introduction of clothing. Lice are highly specialised when it comes to selecting their habitat. Nowadays, every ape has its own species of louse, but humans can have up to three, one for the hair, one for the private parts and one for clothing. This fact turns out to be very useful for our purposes.

Analysing the different DNA structures of our parasites tells us that the common ancestor of the chimpanzee louse and that present in human hair dates back to 6 million years ago, confirming the date when our evolutionary lines diverged. Since then, hominins and chimpanzees have evolved in harmony with their respective lice.

At a certain point, the hominins nonetheless acquired a second species of lice that lived in the pubic hair (commonly called crabs), which evolved from those of the gorilla 3–4 million years ago. Glossing quickly over the embarrassing question of how this transfer could have occurred, this finding suggests that the reduction in human body hair must have started during that period. In order for two different species of parasites to have developed, the follicular habitats on the head and pubic area must have been separate. This means that Lucy was probably already "naked", or almost. In any case, her abdomen must have been sufficiently hairless to enable two different species of lice to evolve.

So when did we begin to wear clothing? Humans have a third body louse that lives on their clothes (where it lays its eggs). From there, it moves onto the skin several times a day to feed. Genetic tests based on mitochondrial DNA mutations tell us that the louse that lives on our clothing separated from the head louse between 83,000 and 170,000 years ago. This means that we might have started to wear clothes in Africa before our exodus from that continent.

The reason why remains a mystery, given that African humans could easily sleep naked, possibly heated by fires arranged around their couches on cooler nights, as some peoples still do even today, weather permitting. We also do not know if and how the other archaic humans, now extinct, dressed during glacial periods in Eurasia. One would perhaps assume that the latest species of lice evolved on the

Fig. 6.1 Artistic representation of hominids' biodiversity. *Source*: Drawing by Tullio Perentin, ZOIC, Trieste

clothing of the Neanderthals and Denisovans, who lived in colder climates. Later, they could have survived on our own clothing due to the encounters that we undoubtedly had with those other species.

There is also a suggestion that, just as evolution selected skin colour based on exposure to the sun, the hairy pelt could have become more sparse or thicker depending on the temperature, the level of environmental humidity and the number of available sweat glands.

6.2 Clothing, Roles and Images

Later on, we will discuss another possible explanation for why we started to wear clothes at some point, irrespective of environmental demands. For the time being, suffice it to say that dressing could also depend on the image we wish to convey of ourselves to other members of the society. By dressing, we interact socially and can express an idea of who we are. Not who we really are (this is our secret, of which we are sometimes unaware), but rather who we wish to be perceived as. Ultimately, if this artifice works, the fictitious character becomes real, capable of generating roles, hierarchies and relationships, giving rise to meanings that we ourselves have created. The identification of our own individual identities with an artificially constructed image is one of the many ways of connecting socially and bringing about significant effects, as we shall see.

Going back to our story, when the *sapiens* emigrated from Africa, they carried with them both their parasites and the new innovation of clothing. This must have proved very useful when dealing with the icy steppes of Eurasia. They would have been able to use the pelts of various animals, including those of *Ursus spelaeus*, the cave bears that became extinct at the end of the last ice age.

US researchers have shown that European and Chinese *sapiens* from 40,000 years ago had already invented shoes. This can be deduced from an analysis of their toes, which became weaker with the regular use of footwear. Because the foot bones of the Neanderthals do not display this feature, this means that either they possessed shoes that were less comfortable or that they went through all the ice ages barefoot!

We have recently discovered other details, including the colour of their residual hair. Sequencing of Neanderthal DNA has made it possible to identify a variant of the MC1R gene (which encodes a protein used to produce melanin), which is associated with red hair on the head and body, as well as fair, freckled skin.

Whether or not they wore shoes, their long survival through the various ice ages suggests that the Neanderthals must definitely have developed techniques for producing clothing out of animal skins. We have already mentioned their invention of bone tools for making the skins waterproof. Now, we also have evidence that our direct ancestors also produced pins, needles and other tools for making increasingly elaborate garments. They were evidently not prepared to wander the Eurasian steppes barefoot and poorly dressed. This was to prove particularly useful later on when crossing the icy wastes of Beringia to reach the Americas.

Chapter 7
Lucy and the Other Ladies

So far, we have often spoken about men, but what happened to all the women? Can we blame it all on Carl Linnaeus, the Swedish naturalist who coined the term *Homo* in the eighteenth century to describe both sexes? The term is somewhat inapt for describing a female. We must make an effort; we need to be more specific. How many of us would immediately conjure up an image of a female on hearing the term *Homo erectus*?

Narratives describing our origins almost always refer to a male hominin, who evolved through the development of brains and dexterity. Women remain in the shadows, although sometimes we emphasise the fact that they had to deal with an increasingly risky childbirth and found themselves having to care for children with longer and longer childhoods. In the meantime, males improved their stone tools, making them ever more deadly for hunting and fighting (a typically male function). Yet, these tools could also be used for foraging and then harvesting and cutting up products found in nature (a typically female function). The cave paintings that adorn thousands of caves and shelters of the past with images of large animals are also usually ascribed to male artists, but is this assumption true? So, let us now concentrate, for once, on women, using this term to include the female archaic hominins, who are well represented in the fossil record.

The most famous finds are, indeed, often of women. For example, the *Ardipithecus ramidus* individual Ardi, the *Au. afarensis* skeleton Lucy and one of the recent *Australopithecus sediba* individuals, as well as the first Hobbit to emerge from the excavations in the Liang Bua cave, were all female. The Denisovan *Homo* is also a woman, as are many other *erectus*, *neanderthalensis* and *sapiens* finds. We must, therefore, use the term *Homo* with great care; we never know whether it is being used to describe a man or a woman in any given case. How is it possible to establish gender from a few fossilised bones?

In recent years, we have been able to carry out DNA analysis routinely, and this is now being extended to increasingly ancient fossils that, in the past, could only be examined by means of morphological analysis (which is still a good method). In adult hominins, particularly the most archaic ones, females had a much smaller structure than males. Today, these differences are very minor in our own species, but in those days, they were often pronounced. Specific bones also had different morphologies in males and females, especially the pelvis. In the female, the pubic arch is wider than in the male. Her bones are also generally thinner and lighter.

These characteristics were found in Lucy, the Australopithecus specimen we have already spoken about. This find was doubly lucky: not just an almost complete skeleton that could be subjected to a morphologically accurate analysis but also one whose remains were trapped in layers of volcanic ash that allowed us to use a dating method based on the decay of potassium-40. This nuclear clock told us that Lucy lived 3.2 million years ago. She was 20 years old, measured 110 cm in height, weighed 30 kg, had a brain with a volume of approximately 400 ml and walked upright. Her remains help us to glean important details about evolution.

7.1 The Obstetric Dilemma

Australopithecines had a wider and more concave hipbone than that of the chimpanzees. The pelvic adaptation to bipedal locomotion was achieved by means of the muscular reorganisation necessary to keep the body balanced in an upright position. While the new pelvic structure facilitated bipedal locomotion, it unfortunately increased the risks involved in childbearing. To explain this, we will try to draw some comparisons between different species.

A small chimpanzee can pass through the birth canal without any rotation, emerging with his neck facing backward and his eyes facing his mother. His head is relatively small in relation to the size of the pelvis. During childbirth, the mother does not, therefore, generally risk either her own life or that of her offspring.

In the australopithecines, the baby needed to turn around at the beginning of the delivery process, but could then descend along the entire birth canal without further rotations. Childbirth began to be risky, but not excessively so.

One million years later, the female *ergaster* had a pelvis that required a much more complex delivery process, given that she was adapted to better locomotion but now had to give birth to children with a larger brain. Childbirth became more challenging.

Birth became even more difficult for the small *sapiens* with even bigger heads. First of all, the shoulders had to be aligned with the main axis of the canal on the way in and then turned again to line up with its main axis on the way out. At the end, the baby has to come out with the back of its head turned toward the mother after going through two half turns. The benefits of the upright position were thus offset by more and more disadvantages that were leading us towards an increasing rate of mortality among women and babies.

As the various human species evolved, we found a very effective solution to the obstetric dilemma described above.

We could not change the female pelvis (or at least, not very much), since this remained the pivot we relied upon for our upright posture and balance during bipedal locomotion for both genders. Though the female *Homo* admittedly had broader hips than her ancestors, not many more tweaks could be made to sexual dimorphism: females could not be hindered from walking efficiently in order to facilitate the reproductive needs of the species. It was, however, possible to change the timing of the birth, making it take place earlier. In this way, the full development of the baby's head could take place partly in the uterus and partly after birth. This is exactly what seems to have happened. If we make a comparison with the world of all the other primates, it turns out that we *sapiens* are an exception in at least two respects: (i) the large size of our brains in relation to our bodies, and (ii) the shorter duration of our gestation. Let's take a closer look at this second point.

A typical non-human primate with a brain as big as our own should by rights have a gestation period of 18 months (in order to be in line with the development of the other apes), whereas our pregnancies actually last only nine months. Furthermore, the first months of life are characterised by the infant's total inability to do anything for itself; indeed, the baby lacks even the most minimal survival functions. During the first year or so, the brain (and the nervous system in general) continues its development. This is made possible due to the presence of non-ossified parts in the skulls of newborns such as the "fontanelles", which were already present in archaic hominins and also serve the purpose of making the skull elastic, facilitating delivery.

This practically amounts to completing our growth (specifically that of our brain) in a phase of life that is still foetal but is conducted outside the womb. Now, everything begins to add up: the first part of the 18-month ideal gestation period of the large-brained primate (human) takes place in the womb and the second part takes place outside it. The largest possible newborn head size that will allow the baby to be born alive without killing the mother has been calculated, amounting to 500 ml. The *sapiens* newborn keeps below that limit with a volume of about 350 ml.

The shortening of pregnancies may already have begun in the prehuman hominins. Recent studies on an *Au. afarensis* girl discovered in Dikika, Ethiopia, who died at approximately 3 years of age 3.3 million years ago, show that she was already born prematurely compared to chimpanzees. Her brain, in fact, measured only 330 ml, corresponding to approximately 70 % of that of an adult, while at that age, the brain of a chimpanzee is equivalent to 90 % of an adult brain.

The solution of making the birth happen earlier, achieved by naturally selecting for premature birth, actually comes at a minimal cost, which is mainly borne by the mother; she simply needs to look after premature babies for longer. This is definitely an acceptable sacrifice and perhaps not a sacrifice at all, when compared with that of risking her life together with that of her descendants, or of not being able to walk swiftly.

7.2 Turning a Disadvantage into an Advantage

With *sapiens*, the stratagem of modulating the timing of maturation extends not only to the postnatal period but also throughout childhood as a whole, all the way up to adolescence. Nowadays, we are the only primates to have such a long relationship with our descendants. Why should that be?

Some maintain that this is an evolutionary advantage. A brain growing outside the womb is able to absorb an enormous amount of information and is very receptive in social and environmental terms. It is, therefore, malleable with regard to the information and skills that require enhancing. We will see this advantage prove to be crucial in the transmission of culture through generations and in the development of cognitive abilities.

Even the development of language (spoken and sung) could be attributed, at least in part, to the relationship between mother and child due to the fact that a hairless mother is meant to hold her child in her arms. Because physical contact is so imperative in primates for a more insecure and immature young, vocal reassurances from mother to child may have reduced the demand for this contact, allowing the mother to set her baby down and have greater freedom of movement to perform her daily activities. Through natural selection, this ability could then have been handed down and enhanced by adding in all other types of verbal communication, even between adult individuals, duly supported by the development of the relevant areas of the brain. The hypothesis that women are responsible for paving the way for the "invention" of language, and perhaps even for the development of the first tools, seems to be borne out by the observation that in societies that are still promiscuous and, therefore, lack a father figure, the job of rearing offspring falls predominantly on mothers, who thus become a primary source for transmitting knowledge, or at least mould the children's cognitive capacity.

Recently, we have discovered interesting details about the relative growth of children's brains in Neanderthal and *sapiens*. Immediately after birth, the brains of the two species are very similar in shape and size (although the Neanderthal brain is slightly elongated at the rear). During the first year of development, the structure of the skull is not totally ossified and can still change its shape. At one point, we and the Neanderthals start to diverge. In particular, *sapiens* goes on to develop a skull shape that would make room for those parts of the brain that are crucial for visual and spatial coordination, as well as for the development of symbolic thought. We shall argue that this latter feature, although not totally absent in the Neanderthals, turned out to be of paramount importance for our survival as a species.

7.3 Teeth: The Black Box of Our Lives

Going back to the moment of birth, science allows us to obtain very useful information when teeth are found at archaeological sites. All too often, they are the only human remains that are left. We can read in them the daily and weekly growth lines of the enamel, which are produced by the biological cycles that characterise the cellular secretion. In particular, episodes of stress remain etched forever in the growth layers of tooth enamel. Teeth are indeed a kind of black box recording the experiences of their owners.

In our case, they are visible in mothers, as well as in babies. In women, they reveal the number of pregnancies and any birth difficulties. In babies, they reveal the stress they had to endure in order to be born. It is even possible to determine the age at which they were weaned or at which they died. Information on mothers can be obtained by analysing the growth layers of the third molars: these teeth, known as wisdom teeth, emerge during their early childbearing years.

Synchrotrons, radiation sources developed for very different purposes, can be used to acquire images of the internal structure of fossilised teeth without damaging them. A synchrotron produces high intensity x-rays by accelerating electrons at speeds comparable to that of light while they move along a circular path inside a doughnut-shaped ring that is maintained under a high vacuum. The information gained with synchrotron CT scanning is very detailed. It reveals the developmental age of a hominin, confirming what we said earlier about the progressive lengthening of the *Homo* childhood and adolescence along his evolutionary line.

The enamel microstructure tells us that all developmental stages were shorter for australopithecines than for humans; indeed, they were still similar to those of present-day chimpanzees. Modern humans take twice as long as the chimpanzees (and probably also the australopithecines) to reach maturity. The growth period of the first *Homo* species was probably halfway between that of modern humans and that of present-day apes. Childhood began to lengthen only when human evolution was dominated by brain growth. Both aspects are actually related and critical to our evolution as *sapiens*: a prolonged childhood makes it possible to exploit our new brainpower more efficiently and increases the ability to learn and convey knowledge.

What did we learn from the x-ray imaging of Neanderthal children's teeth? It seems that their postnatal development was shorter than our own. It is, therefore, justifiable to wonder how important this aspect was in reducing their learning potential compared to their *sapiens* peers and whether there is any chance that this was also one of the factors that reduced their ability to compete with us, leading them to extinction a few thousand years after our arrival.

Fig. 7.1 Recent reconstruction of Neanderthal woman with children. *Source*: By kind permission of P. Plailly/E. Daynés-Reconstitutions Atelier Daynes, Paris

7.4 "Multimedia" Ceremonies

The *sapiens* women continued to care for their young, who went through increasingly long childhoods. This made the women less mobile. We wonder whether it was the women who left us those marvellous wall paintings of animals—bears, rhinoceroses, lions and panthers—in the Chauvet caves in France. After all, it is possible that the large animals could have been drawn not only because they were hunted, such as the horses, reindeer and deer drawn thousands of years later in the Lascaux cave, but also because they could stir strong feelings and allow for the performance of large ceremonies.

In any case, whatever the gender of the artists and the purpose of the paintings, the depiction of the different animals in the Chauvet cave, in which the heads and legs of each figure are drawn several times in a sequence, side-by-side, with profiles of animals of different sizes, has something extraordinary about it. Some have

7.4 "Multimedia" Ceremonies

Fig. 7.2 Rock art in Chauvet, France. *Source*: By kind permission of Jean Clottes

argued that this marked the first attempt at dynamically representing observed reality, a kind of show before its time. The cave, which boasts an area of 36,000 square metres, a curved "screen" and 442 animals drawn on the walls (together with explicit female references), could well be the first known example of an Imax cathedral, in which a sort of ceremonial "movie" could be performed. An experiment has shown that a sense of movement would be achieved by combining this painting technique with the effect of flickering torchlight in a dark environment, making the viewer feel centre-stage, surrounded by a great deal of action. We are sure that such a show, depicting hunting scenes between animals, perhaps accompanied by a story told or sung, would have stirred deep emotions in the audience of those times. The same feelings are aroused in contemporary humans today in a digital reconstruction of the Chauvet cave, in which a computer simulates a replica of the paintings under flickering lights, allowing the viewers to experience the jaw-dropping sight of those animal images in movement.

Chapter 8
Menus of the Past

Today, there is much talk of what foods suit us. Perhaps out of shame for our ongoing mass exploitation of the animal kingdom and fear of the effects of a diet based on intensive stock-rearing and farming, some of us take refuge in vegetarian cuisine, even better if it is organic.

It is hard to deny that a carnivorous diet such as that prevailing in developed countries is no longer sustainable for the planet if we all choose this path. Are we really bound to do so? What did hominins eat in the deep time of our evolution? This is not a matter of idle historical curiosity. After all, we are not only "what we eat"; we are also "what our ancestors ate".

The ancestor that we shared with chimpanzees some 6 million years ago would probably not have turned down meat, which is also occasionally eaten by the great apes of today. Even *Au. afarensis* ate some meat. In 2010, the bones of animals possibly butchered by these hominins using fairly primitive stone tools were found in Ethiopia in the geological record dating back more than 3 million years. However, they generally preferred to feed on fruits and vegetables. These latter dietary preferences may have led them to extinction when climate changes dramatically altered their habitat. The last australopithecines, such as *Australopithecus sediba*, dating back about 2 million years, had resigned themselves to a diet of leaves and tree bark, as may be deduced from the carbon isotope analysis of the phytoliths contained in their dental tartar. Phytoliths are microscopic minerals that form in cells of many plants, remaining well-preserved for long periods after the death and decay of the plants. On the other hand, recent biomechanical analyses show that *sediba* had a very weak jaw, not adapted to a diet based on hard foods. His fate was sealed.

Until recently, it was believed that the African forests thinned out slowly and gradually over the last few million years, giving way to the savannah. In actual fact, the forests did not exactly disappear in this way. Recent analysis of marine sediment cores taken from the Gulf of Aden near the African coast shows that the periods between 2.9 and 2.4 million years ago and between 1.9 and 1.6 million years ago correspond to particularly intense climate and environment changes, characterised by a rapid succession of forestation and deforestation. Each of these periods corresponded, in turn, to a significant step in the history of human evolution. The first period saw the appearance of the first human forms, such as *Homo habilis* and the 2.8-million-year-old *Homo* recently found in Ethiopia, while the second saw the appearance of *H. ergaster*, the first definitively human hominin.

Different evolutionary responses to environmental variations can be observed. On the one hand, some species adapted through morphological changes more consistent with the newly available natural resources (for example, by means of different chewing mechanisms). On the other hand, new solutions began to be introduced, such as an increase in brain size, with particular emphasis on the development of the neocortex and frontal lobes. It was as if biological evolution had become too slow for certain groups of hominins and it became necessary to develop the ability to respond faster—through changes in behaviour, learning and socialisation—to the effects of the most rapid climate fluctuations. Due to the energy costs of a larger brain—even though the brain represents only 2 % of body mass, it consumes at least 20 % of the total energy needed by the body—this option became more viable only when its evolutionary benefits outweighed the higher energy consumption. It was at this point that a very significant gap began to open up between the different species of hominins, something that was to affect the course of our evolutionary line decisively. Here are some examples.

Australopithecus boisei, whose first fossilised remains date back to approximately 2 million years ago, was named Nutcracker Man due to the fact that his molars, along with having thick enamel and being set in stronger jaws, were flatter than those of the first australopithecines. These hominins, nowadays preferentially called *Paranthropus*, had powerful chewing muscles. Isotopic analysis of the tooth enamel confirms a change of diet, from fruit and leaves to berries, roots and tubers. It shows that they also ate termites. A similar hominin, *Paranthropus robustus*, appeared simultaneously in Southern Africa. During the same period, the structure of the small brain of *Australopithecus sediba* had begun to change, developing some of the first features that would later become typical of the human brain. *Homo habilis*, whose oldest remains date back to slightly earlier than 2 million years ago, had less pronounced teeth and jaws but a much larger brain, measuring more than 700 ml. He was fully bipedal, with shorter fingers and toes that were not very useful for climbing trees. He gained his name from being considered the inventor of the first stone tools, even though recent discoveries challenge this distinction.

In the end, the evolutionary line based on adaptation of the chewing apparatus died out: all *Paranthropus* disappeared approximately 1 million years ago. The adaptation based on changes in the brain, contrastingly, continued successfully.

Homo ergaster, with his 900-ml brain, appeared in Africa approximately 1.9 million years ago. This hominin had smaller jaw muscles and molars, suggesting a diet of soft foods. As we said, *ergaster* was able to cook meat and could ingest large quantities of protein that took less time to digest, thus further changing his anatomy, including his rib cage and hip structure. Turkana Boy, a young *ergaster* found near Lake Turkana in Kenya, lived approximately 1.6 million years ago and, by the age of nine, had already reached a height of nearly 160 cm, twice as tall as some

Australopithecus. His age at death and biological development were determined through analysis of his teeth by means of a CT scan with a synchrotron light source.

The brain of *ergaster* was more than twice the size of that of the first bipedal apes and it could continue to grow, also thanks to his diet of cooked meat. Some australopithecines and even individuals from their own species probably ended up on the menu of these early humans.

8.1 Ritual Food

Evidence that our ancestors might have been cannibals comes, for example, from the Atapuerca site in Spain, where unmistakable signs of butchery were found on the bones of archaic humans from about 800,000 years ago. This practice was continued by the Neanderthals and modern *sapiens*, some of whom were known to practice it as recently as a few decades ago, although associated with ritualistic behaviour rather than an actual need for nourishment.

This prevalence of ritual over nutritional purposes seems to be confirmed by observing certain peoples who, until recently, kept up this practice and only subsequently abandoned it for health reasons. As an example, Papua New Guinea banned cannibalism in the mid-1950s. Yet, this was not done for ethical reasons, but rather to eradicate kuru, a neurological disease caused by eating human brain during funeral rites.

The act of eating can also have at least two antithetical meanings: on the one hand, it expresses "communion", while on the other hand, it expresses "aggression". Unsurprisingly, the former meaning was sometimes translated in transcendent terms, accompanying rituals that persist in a sublimated form to our own time, whereas the latter became a taboo: it is forbidden to eat your own kind because they are members of the great "human family". Over time, rules and customs of the latter type would be extended to certain animals performing family functions, for example, animals used for company, security, hunting or transport (such as dogs, cats and horses).

At a certain point, for us *sapiens*, food began to be linked to the passing on of legends and religions, and even the creation of institutions regulating social life. We need only think of rituals involving animal and even human sacrifices that are widely documented in many cultures. These forms of behaviour are associated with the construction of imaginary realities, the result of new and specific mental abilities. The evocative role of food persists in modern society, as can be seen from the rituals performed every day in our own homes and in restaurants. Business lunches, family meals, wedding banquets and company dinners are all examples of institutionalised occasions mediated by food and providing an opportunity to reinforce the rules of coexistence. Even the dietary rules imposed by different religions are innumerable, embracing Halal food, Kosher food, Lent and Ramadan, taboos on eating certain animal species and the custom of eating others on certain occasions. Food is also often a symbol in itself: we need only think of the bread and wine associated with the Christian religion. Lastly, our obsession with becoming thinner (or fatter) can be traced back to other myths such as "perfect beauty" or "perfect health".

8.2 Vegetarians or Carnivores? Omnivores!

The Neanderthals have enjoyed the notoriety of being insatiable carnivores since their discovery one and a half centuries ago; many animal bones have been found at certain of their sites along with Mousterian stone tools. Isotopic analyses of their bones initially seemed to confirm this, but recent tests on their teeth indicate that some of them followed a more varied diet, which included tubers and other tough vegetables. An analysis of phytoliths and starch in dental tartar, a reliable archive of dietary habits, also tells us that they were eating and sometimes cooking different kinds of vegetables. Surprisingly enough, an analysis of tooth tartar from some Neanderthals found at the El Sidrón site in Spain showed no trace of protein derived from meat consumption. Yet, it is hard to believe that the Neanderthals were generally vegetarians.

Recent analyses of sediments containing the remains of fossilised Neanderthal faeces from 50,000 years ago, discovered at El Salt, Spain, support the theory that their diet was partly based on meat consumption. In other cases, foods such as chamomile and yarrow, which are known for their medicinal properties, have also turned up in their diet. A gene known as TAS2R38 has been identified in the Neanderthal DNA. This makes it possible to perceive bitterness and is present in our own genome to help us detect the presence of toxins in plants. It is, therefore, reasonable to conclude that the Neanderthals were omnivores.

What do we know of our *sapiens* diet back in the days when we were hunters and gatherers, in other words, before the advent of farming? This is a serious question. Our current dietary behaviours are, of course, influenced by what has happened in the last 10,000 years, but also by the eating habits that characterised the other 190,000 years of our history as a species. We know that our diet was wider and more varied throughout this long period than at present. Despite the importance that we attach to the hunting of large animals, which required a strong spirit of cooperation and coordination between hunters, most of our diet was based on gathering produce from the land and catching small animals: fish, shellfish, grubs and insects.

Sugary food such as good ripe fruit was rare to come by and difficult to store. This meant we had to eat as much as we could as quickly as we could before anyone else came along to steal it. We still feel this temptation deep inside when we look at the cakes on display on the shelves of a pastry shop. In our subconscious, we are still living in the savannah and often struggle to curb our greed. Diabetes, obesity and heart disease may be contemporary global problems but they have very deep roots.

8.2 Vegetarians or Carnivores? Omnivores!

Before farming, a more varied diet, little sugar and a lot more exercise meant that we were presumably healthier, stronger and more athletic than today. Our skills must have been broad and applicable in many fields, as we could generally only rely on ourselves and a small band of companions.

We also had to be familiar with the features of the land we inhabited, sources of water, and food-gathering and prey-catching techniques, as well as what was edible and what was not. If any contemporary human tried to compete with our ancestors on a hypothetical trip back in time, it is very likely that he/she would find it much more difficult to survive despite all the knowledge available to him/her. Our hopes of survival would increase only if we were a good-sized group of individuals, each with different skills and abilities. We have chosen to specialise more and more in a few areas. By doing so, we have a very high pool of resources to draw on, but only as members of a society. Our knowledge is much broader as a community, but much more limited as individuals. We will return to this point later.

At the end of the last glacial period, we know that *sapiens* developed agriculture and selectively bred certain animals for food. With the increased availability of cereals and meat, women were able to procreate every two and a half years, giving them an edge of approximately one year over women belonging to a society of hunter-gatherers. This phenomenon produced an incredible population explosion, only partly mitigated by wars, famines and pandemics, which wiped out almost all previous hunter-gatherer populations.

Contrary to popular belief, even though farming and stock-keeping increased the amount of food available, the variety increased only for the wealthiest. Even today, the bulk of humanity lives mainly on a few staple cereals. In general, the diet of farmers and sedentary stock-keepers opened the way to new diseases: tooth decay, periodontitis, infectious diseases, and deficiency in iron and other vital elements, as well as the spread of intestinal parasites.

Nowadays, after 10,000 years of experience and with lower availability of many natural sources of protein (we need only think of the impoverishment of the sea), we are beginning to feel the limits of this development, which seems inefficient and unsustainable given the increasing scarcity of resources. If we continue to grow at our current rate and do not wish to become totally vegetarian, a much more promising approach would be to adopt a diet based on smaller, more abundant animals, including insects and grubs, whose populations are still expanding like our own and, when fried, constitute part of the diets of many people today (particularly in Asia and along the Tropics).

Few people are probably aware that, according to a recent report by the Food and Agriculture Organization of the United Nations, a diet of this kind is currently followed by 2 billion people: more than one quarter of the world's population. In particular, among the 2000 species of edible insects, cockroaches are the most popular (31%), followed by caterpillars (18%), ants, wasps and bees (15%), and crickets and locusts (13%). The rest is divided between dragonflies, flies and termites. Their nutritional value is actually no lower than that of meat. Their environmental impact, in terms of greenhouse gas production and land use, is also more than four times less than that of cattle, pigs and other animals that we have been raising for 10,000 years. By following a diet of this type, it is also possible to convert low-value organic waste (on which insects feed) into proteins.

To meet the tastes of industrialised countries, projects are currently going ahead to market these "energy sources" through the production of flours that can be used in the various dishes people are accustomed to eat in the developed world and make them more "appealing" to Western eyes and taste. Recently, China has begun a series of tests to establish the possibility of including certain grubs (including silkworms) as food during space travel. To begin with, three people were locked away for 105 days in a laboratory and fed on a diet based on a beetle, *Tenebrio molitor*, which was reared on plants grown in a "bioregenerative life-support system". The renowned Chinese talent for spices will undoubtedly prove very handy when it comes to making a diet of this kind more palatable.

Fig. 8.1 Fried Ensiferum (Orthoptera), a nutritious diet

8.3 Us and Them: Cain's Diet

So far, we have talked about our attitude towards food and how we have learned to change our eating habits to adapt to the availability of food. We can now look at things the other way around and determine whether, apart from diseases, food has had any influence on us, not only in terms of our physical features (height, strength, agility, brain mass and teeth) but also in terms of our individual and collective behaviour. We even wonder whether our eating habits could have had an impact on how we think and how we relate to one another.

If we look at our current behaviour, we know that we often develop hostile attitudes towards "others". But who are these individuals? And how do we distinguish between "ourselves" and "the others"? In most cases, "the others" are individuals who we think are different because they speak incomprehensible languages and have an unusual appearance and habits that are alien to us. We then become particularly aggressive when we fear that they will create economic and social problems, threatening our way of life. The diversities that annoy us can relate to skin colour, facial features, social status, sexual preferences, dietary preferences, political ideologies, religious beliefs and more.

Human history is peppered with bitter conflicts between groups that consider each other to be "different". On the other hand, we also sometimes show a propensity for cooperation and empathy, helping those in need or teaming up with those who share some characteristics with us (trivial examples may include fans of a particular team and those who share the same taste in dressing or entertainment). The boundary between "us" and "them" changes all the time and could be that of your own neighbourhood, your own nation, your own ethnicity, and your own religion—even your own gender. It has always been believed that these distinctions arise out of cultural, social and economic legacies and that we can fight them off given a certain amount of goodwill and through the use of reason. The task would be much more difficult if it were to prove that this shifting relational structure is hard-wired into us due to our genetic make-up and the way our brain works.

Very recent studies suggest that this tendency towards "social dualism" (the arbitrary and variable inclusion of some individuals in our communities to the exclusion of others) is one of the main traits responsible for our original success as a species and was determined by a radical change in our eating habits. This behaviour could unfortunately backfire within our present-day society. Let us look at why.

We know that 150,000 years ago saw the beginning of a period characterised by increasingly intense climate change. Conditions were tough for everyone, including the different human species that populated the planet at that time and the animals and plants that made up our diet. The different human species did not have much opportunity to come into conflict with one another, because they lived quite separately in geographical terms. Some anthropologists suggest that under these conditions there was scope for conflicts within our own species. What could have been the reason for this?

While the cooler Eurasian climate placed the Neanderthals under stress, the dry spells recorded in Africa made the savannah increasingly inhospitable for us *sapiens*. Recent research has nevertheless revealed that the coastal environments of Africa brought salvation for our direct ancestors who, by then, had dwindled greatly in number. The coasts, being packed with seafood and other localised food supplies, offered stable and predictable resources that were particularly valuable given the scarcity of food. Under such circumstances, it became necessary for the first time to form groups to defend the territory in order to survive and thrive. Inland, it was not particularly helpful to coordinate large groups because the food was scattered, mobile and unpredictable. Alliances were limited to certain big hunting expeditions, but it was normally too costly to oversee large areas in order to keep others out. On the other hand, it is estimated that along the coast an individual could easily gather 5000 calories' worth of shellfish in one hour. This new prey was unable to escape and was, therefore, highly desirable. And so, for the first time, it was appropriate to invest in protecting those resources, depriving others of access. This marked the creation of the first private property, albeit group-owned. All that was needed was to identify a criterion used to define membership of a particular group through family ties, kinship, or recognisable external "markings".

This situation would, of course, have paved the way for a better understanding between members of the same group, but also for harsher conflicts between different groups. The outcome would have been the first wars over territory, events that would have been pointless in large environments with abundant, mobile resources. Evidence on such events is hard to find though.

A great number of human skeletons (men, women and children), showing traumatic lesions produced by lithic weapons, were recently unearthed at a 10,000-year-old site near Lake Turkana, in northern Kenya. The area was then a fertile lagoon that could sustain a large community. This is the smoking gun confirming that violent warfare was part of the intergroup relations among hunter-gatherers. Yet, we are inclined to believe that such wars must have taken place for tens of thousands of years, long before the wars that historians believe were first waged over agricultural resources (which lay at the basis of the known ancient civilisations) and subsequently over communication pathways at certain strategic points of the territory. They would have also given strategic advantages to those groups who were better trained and equipped for combat and had better weapons, marking the beginning of the arms race.

In due time, differences among groups could be fostered and maintained through language variations. The concentration of physical characteristics, such as skin and hair colour, eye shape, and body structure—when easily recognisable—also helped to distinguish between "us" and "them". Today, we are all too familiar with these topics.

If this is true, for tens of thousands of years, natural selection has been favouring individuals with genes and brain connections fostering cooperation within groups of like-minded people and competition with different groups. This could well be the explanation for our propensity to create social groupings based on (arbitrary) categories of homogeneity and diversity. This also explains the ambivalence and variability of our current cooperative and competitive nature. If this characteristic is really part of our genetic heritage, it could not have arisen a few thousand years ago with the advent of agriculture. The timeframe would have been too short.

Nowadays, we have other resources to be defended or conquered: accumulated and poorly distributed wealth, black gold (oil), blue gold (water), markets, technologies, the rights of citizens and their lifestyles. There are certainly plenty of reasons for conflict. If our "territorialism" were only a cultural trait, there would be some way to get around it, but if it really is a genetic feature that has developed over the last hundred thousand years, it is much more difficult to envisage a solution that is "good for all", in an overpopulated world endowed with increasingly scarce resources.

8.3 Us and Them: Cain's Diet

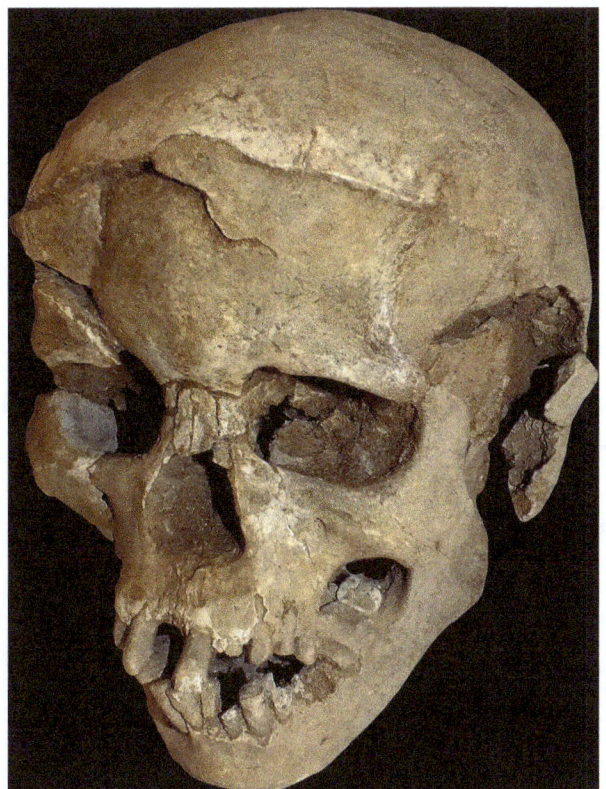

Fig. 8.2 Skull with multiple lesions, consistent with wounds from a blunt instrument (10,000 years ago). *Sourcen:* By kind permission of Marta Mirazon Lahr

Chapter 9
Ancient Ills and Ancient Remedies

Do you have cirrhosis of the liver or diabetes? Are you finding it difficult to kick a habit? Are you depressed? Do you have skin cancer or incontinence? The blame may not lie with you or your lifestyle: you may have a genetic predisposition inherited from the Neanderthals.

We have already mentioned that Neanderthal and *Homo sapiens* individuals interbred on various occasions. The traces of those ancient relationships have naturally become diluted over time and now account for 1 to 4 % of our DNA. The regions of our genome that are derived from the Neanderthal contribution contain a few hundred thousand bases at most (out of the 3.2 billion base pairs that make up the human genome). Yet, because Neanderthal DNA differs from our own DNA at specific locations of the genome, those sequences can be used to identify its genetic contribution. The Neanderthal contribution has been directly confirmed in the genomes of more than 1000 present-day humans.

In consideration of the genes specifically associated with our state of health, some US scientists had discovered that DNA of Neanderthal origin might be linked to a predisposition to diseases such as diabetes, Crohn's disease (a form of bowel inflammation), lupus (an autoimmune disease), cirrhosis of the liver and difficulty in giving up dependencies. This negative contribution is offset by genetic influences conferring certain advantages, such as those connected with the presence of keratin, which gives us skin, hair and nails that are more impermeable and resistant to the cold. It could generally be said that these crossbreeding processes accelerated our adaptation to a different environment but also left us with some weaknesses.

The above hypotheses have been recently studied by comparing the medical record of 28,000 adults of European ancestry with the Neanderthal gene variants of the individuals in the database. It is confirmed that our Neanderthal genes increase the risk of dermatological, immunological, neurological, psychiatric and other disorders. This does not mean that the Neanderthal and the *sapiens* were suffering from these diseases. Some of the Neanderthal genes might have provided a benefit for the early African *Homo sapiens* migrating to Eurasia, but they are a health risk for the present-day humans, particularly those living in industrialised societies. Genes that strengthen our immune system were probably very useful during the last ice age, but today they increase the risks of inflammation and allergies. Hypercoagulation was of paramount importance for the Palaeolithic lifestyle, but today it would increase the probability of strokes and other disorders relate to blood clots.

Other areas of our genome are nevertheless completely devoid of any genetic contribution from the Neanderthals. This means that other changes that could have been introduced into our genetic make-up have probably already been removed by natural selection in cases in which they jeopardised our survival. Genes associated with the male reproductive system are one example of an area that is devoid of any Neanderthal contribution. It would appear that the genetic heritage of the Neanderthals rendered male *Homo sapiens* individuals less fertile, lowering their potential

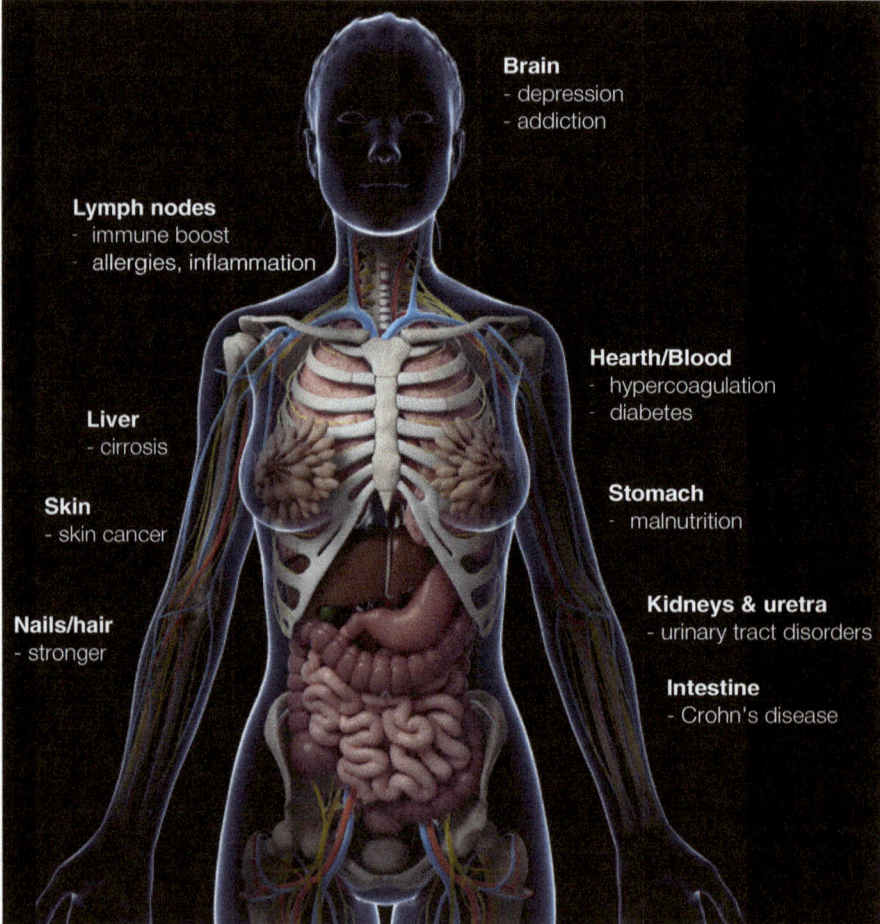

Fig. 9.1 Genetic predispositions inherited from the Neanderthals. *Source*: Shutterstock.com, Copyright: Sebastian Kaulitzki

for generating hybrids. This suggests that when *Homo sapiens* and Neanderthals interbred after 500,000 years of evolutionary separation from their common ancestor, both species were already at the limit of their biological incompatibility. Some traces of this ancient relationship nevertheless remain in the archaeological record. For example, an individual believed to be a hybrid dating back 35,000 years was discovered in the Lessini Mountains near Verona in Italy: despite being a Neanderthal, he sported a pronounced chin like our own.

9.1 Diseases and Treatments

How healthy were the hominins in general? It appears that various diseases, from cancer to tooth decay, afflicted the Neanderthals and the other previous human species, as well as *Homo sapiens*. Cancers were probably rare, given the short life expectancy at that time, which was likely to have been around 30 years. A tumour was nevertheless recently discovered in the rib of a Neanderthal individual found at a site in Krapina, Croatia and dating back 120,000 years.

These fossil remains were discovered in 1899 by the Croatian paleoanthropologist Gorjanović-Kramberger, who studied them using the then recently-discovered technique of x-ray analysis. He found cases of taurodontism, a dental anomaly that gives rise to an enlarged pulp chamber and smaller roots. In our own time, far more powerful methods, such as x-ray microtomography, allow us to observe details in 3-D down to a few thousandths of a millimetre. This very technique was used in 2013 to identify the rib tumour mentioned above.

Micro-CT imaging was also used to study the teeth of an *H. Heidelbergensis* specimen from an Algerian archaeological site dating back 700,000 years. These scans revealed some signs of dental decay involving both enamel and dentin. We had to wait until the end of the ice age for evidence of the first dental treatments. In Northern Italy, a dental caries intervention was identified in the molar of a 14,000-year-old man. In Pakistan, teeth dating back 9000 years were found to have been drilled with a flint bit secured to a wooden stick that was turned with the aid of a bow.

Many finds indeed confirm that teeth were a serious problem for all humans, to the extent that their very survival was at risk above a certain age. Once they could no longer chew, they could easily waste away and become malnourished, unless someone offered to chew their food for them.

In 2012, micro-CT imaging led to the discovery of the oldest known dental filling. The jaw, found in present-day Slovenia at the beginning of the last century and dating back 6500 years, had been on exhibition at the Trieste natural history museum since 1911. Even though it had been already radiographed in the 1920s, nobody had noticed anything in particular until state-of-the-art synchrotron radiation 3D imaging revealed an interesting secret harboured by that ancient Istrian. Someone—whether the individual concerned or an ancient dentist—had used beeswax as material for repairing a damaged canine tooth: the first evidence of a Stone Age filling.

What do we know of treatments administered previously, during the last ice age? We know from the scant fossil evidence that a Neanderthal who lived approximately 70,000 years ago, found in the Shanidar Cave in present-day Kurdistan, Iraq, had survived for a long time without any means of self-sufficiency. This confirms the practice of long-term care by other members of an individual's group. As mentioned previously, traces of medicinal herbs present in the Neanderthal diet appear to confirm that treatments and healers already existed in very remote times.

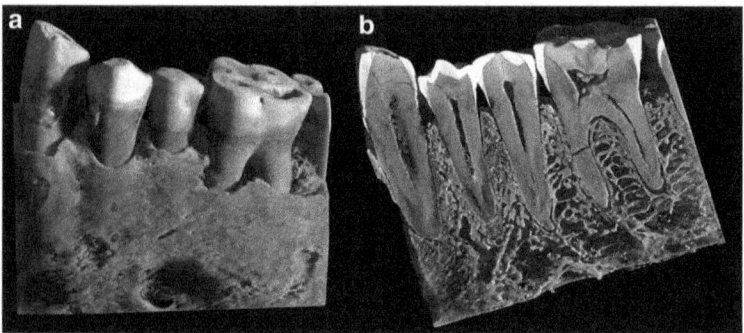

Fig. 9.2 (**a**) and (**b**) Tooth-filling during the Stone Age, discovered through x-ray microtomography. *Source*: Microtomographic reconstruction by F. Bernardini et al. (ICTP, Trieste)

9.2 Placebo Effect

Further evidence of well-established healing practices in the different cultures of our species could help us better understand when and where relationships of trust and empathy began to be built between the different members of a group. Later, we will see that such relationships were to prove crucial in achieving an important step along our evolutionary pathway: the step that led to the emergence of symbolic thought and enabled us to become the social animals we are today. It is now widely accepted that, irrespective of the active ingredients administered, different therapeutic practices are particularly effective when patients are persuaded of the benefits of the treatment and believe in their healer's ability to ward off suffering or death.

This placebo effect has undergone a radical reappraisal in recent years, even by the scientific and medical community. It is not an illusion but, in many cases, a genuine form of therapy with the ability to heal the patient. Historically, the first such practitioners that come to mind are the shamans, who still practise in some communities today. Examples from our own time include the many doctors who deliberately prescribe placebos to accomplish incontestable therapeutic results. Both cases confirm the importance of a positive mental attitude in achieving a tangible physical outcome (healing or even survival). We will see that this characteristic, which is typical of modern humans, will have very many other fields of application. In many cases, the very act of believing allows the desired outcome to be realized.

9.3 Migrations and Contagions

One of our main fears is perhaps that of contracting a disease from people who come from other continents, as we think that our immune system is not prepared to fight it. This fear is generally sound and we are now very fast to set up certain defence measures developed to prevent or deal with such threats.

We know that, during our planetary peregrinations, we have caused the transmission of diseases that decimated or even extinguished entire populations who had never been exposed to them and did not, therefore, have the antibodies they needed to combat them. We have a wealth of documentation dating back 500 years proving that European colonisation was responsible for such events in the Americas and the Pacific.

Without wishing to detract from the overwhelming evidence, it must be said that this view is sometimes untrue. For example, it was initially thought that tuberculosis arrived in the Americas with Europeans a few centuries ago, but we have now discovered that this was not the case. Traces of the disease have actually been identified in a 1000-year-old mummy found in Peru. A recent genetic study tells us that this disease can be traced back to an African population living 6000 years ago. How did it spread from one continent to another? It was discovered that the DNA of the tuberculosis found in the mummy closely resembles that found in whales, which, in turn, closely resembles that found in African goats. In this case, intercontinental transmission of the disease must have occurred due to contact between two different species of animals, which came into contact with humans.

9.4 Self-inflicted Diseases

To sum up this short overview of what we know about our state of health and the treatments available in deep time, we must acknowledge that we cannot blame the Neanderthals for all of our ailments: we have brought many of them on ourselves. By adopting an upright stance, we exposed ourselves to new problems such as hernias, haemorrhoids and varicose veins and we placed our spines under stress, an effect that we feel as we become older. We subjected our joints to stresses and strains by starting to run. In order to speak more clearly, we lowered our trachea and therefore lost the ability to swallow and breathe at the same time, risking suffocation during meals. We exposed ourselves to many diseases by reducing the variety in our diet, thus lowering our immune defences. We have become vulnerable to tooth decay, diabetes and obesity by eating more sugars and cereals. We became settled and packed ourselves into great conurbations where we are more open to epidemics, infections and allergies. By spending many hours reading and looking at screens at close range, we are becoming more near-sighted. By adopting completely unbalanced diets and becoming more sedentary, we build up excess fat and our arteries become clogged. We also often suffer from sleep disorders, which have a negative impact on our day-to-day well-being and also on our mental abilities. When we also consider the pollution that many of us have to live with nowadays, we can conclude that even though our average lifespan has increased, our health has not always improved. Numerous diseases have been eradicated, but others have emerged as a result of a cultural evolution that seems to have outstripped the pace of our biological evolution.

Even though *sapiens* is able to adapt to today's urban and technological environment, he effectively pays a price through increased anxiety and depression and a diminished quality of life, which he often combats with drug abuse. It has also been shown that we are becoming resistant to many ways of treating infections due to the indiscriminate use of antibiotics, directly or through the animals we breed for food. We take a legitimate concern for hygiene to extremes, becoming obsessed with sterilising what we eat and thus eliminating many species of friendly bacteria that help us survive; it has been shown that the effective functioning of our bodily biomes has an impact on our mental health, as well as on our physical well-being.

To sum up, some argue that our culture makes us fall prey to an evolutionary mismatch that leads us to regress because we retain genes that limit our physical and mental efficiency. Many of us already find it difficult to survive without processed food, without shoes, without spectacles, without clothes, without artificial implants, without drugs and even without smartphones and tablets. On the other hand, we increase our chances of survival by bypassing natural selection with ad hoc cultural solutions. In short, we become anatomically more vulnerable, but also stronger and longer-lived due to our inclination to protect those who are genetically weaker.

However appealing the image, the familiar depiction of human evolution as a progressive and linear transformation from an awkward two-legged ape to the elegant human figure we believe ourselves to be is actually somewhat misleading. Our physical make-up—and incidentally that of every other species—is actually the end result of all the evolutionary compromises made by natural selection. This ragbag of components that we have inherited from fish, amphibians, mammals and primates includes recycled solutions adapted from remote times and environments. Some are still useful, some less so. And we are now adding new components, designed by ourselves. Eventually, these will affect the evolution of our minds, as well as of our bodies.

Chapter 10
The Hominin Lifestyle

We have seen that adolescence has become longer and longer during our evolutionary line. We will also talk about how old age has become prolonged. Both are stages in life when opportunities for recreation and leisure increase: we also believe that younger people and older people now enjoy more such opportunities than the young and elderly in the past. How was life for our ancestors in deep time? Is it really true that there was less entertainment and free time? In order to answer these questions, we will begin by taking a look at the length of our ancestors' different stages of life. Then, we will consider some archaeological evidence indicating how they spent their free time.

10.1 Growing Up Too Quickly?

If we look back at the Taung child, the *Australopithecus* specimen from 3 million years ago mentioned earlier, we find that one of his molars had already emerged. Based on *sapiens* biological parameters, this would have made him six years old. However, using tomography with a synchrotron light source, which enables scientists to count the growth lines of tooth enamel, shows that he was only three and a half. The australopithecines progressed quickly from childhood to maturity and their reproductive age occurred earlier than our own.

H. ergaster also developed more rapidly, as can be seen from the remains of the boy from Lake Turkana also previously mentioned. Despite having the appearance of a contemporary 13-year-old, based on his tooth enamel microstructure, we know that he had a biological age of only 9. This means he had a very short childhood, after which he was ready to arm himself with an Acheulean axe and run off into the bush to get a meal. He might have entertained himself by hunting down the last few *Paranthropus*, who eked out a living by eating roots and tubers in the increasingly dry and inhospitable environment.

The young *ergaster* did not remain with his parents for long: his biological clock was ticking and he had to mate as soon as possible to secure a future for his genes. The geological record gives us no clue as to his creativity in terms of arts and tool development. Because his people continued to produce the same stone axe for over one million years, there cannot have been many things to discuss or much knowledge to pass down. So far, there are no hints to suggest that he had already achieved some degree of symbolic thought, although it would have been difficult to preserve evidence of this type of ability in such a remote period, based on a few fossilised bones and stone artefacts. It is likely that their main sources of satisfaction were limited to food, sex and the raising of offspring—an approach still shared by many contemporary humans.

Yet, his *erectus* descendants may have been the first to make the geometric engravings recently spotted on a half-million-year-old shell associated with Dubois' Java Man. According to the discoverers, these signs have a symbolic meaning. This is quite surprising, as it would take us a long way back in time to witness the first appearance of a mental feature that is generally associated with modern humans.

Later humans, such as *H. heidelbergensis*, probably led a very similar life to that described above, or at least we have no evidence to the contrary. Our *sapiens* ancestors nevertheless had a longer childhood, similar to our own, dating back to the time of their evolution in Africa. We know this from our use of synchrotron CT scanning to analyse the teeth of a child from 160,000 years ago discovered in Morocco. He was eight years old and the enamel layers confirmed that his biological development corresponded to what we would expect today at the same age.

10.2 Art and Entertainment

This longer childhood, with no haste to reproduce, proved very useful, because it allowed our brains to mature so that we were able increasingly to learn from adults. We also know that more and more energy was devoted to "entertainment", or at least activities not directly necessary for survival. In Africa, from 100,000 to 80,000 years ago, s*apiens* began to draw symbolic signs on ochre tablets and to create elaborate shell necklaces, as well as developing new, more effective and dangerous stone weapons that they also used for throwing.

Traces of this behaviour, albeit in a different form, are found throughout their expansion eastward, after their exodus from Africa. Handprints (almost certainly female) and animal figures (in this case, the babirussa, or pig-deer, a species of Indonesian pig) dating back 40,000 and 35,000 years, respectively, have recently come to light in some caves in Sulawesi, Indonesia. This discovery pushes back the previous European record for examples of rock art. Further evidence of symbolic thought dating from the same period has been found in Australia.

Once in Europe, the *sapiens* began to represent their world on cave walls. Here, they also began to produce music and artistic or ritual objects. We have already mentioned the Hohle Fels flutes dating back more than 40,000 years. Small headless figurines made out of mammoth ivory with exaggerated female attributes that were designed to be worn around the neck were also found not far away in layers of the same age. Some claim this to be the first example of pornography, but others argue that this view says more about the "observer" than the observed.

Life at that time must have been hard and short as people struggled to live to over 30 years of age, although some of them might have reached the age of 50. When we talk about our ancestors' history, we must remember that we are referring mainly to the history of people whom we would consider young today. Cultural differences aside, there do not seem to be many differences between the ways in which young people liked to entertain themselves then and now. Playing, dancing and singing around the fire in good company would have made it easier to make friends and mate, so as to generate as many offspring as possible. Were it not for the different emphasis on reproductive purpose, which today seems less urgent, there is no reason why our ancestors should not have enjoyed wild parties as much as youngsters do today. After all, we have the same bodies and minds.

In particular, it is interesting to note that one third of our ancestors' lives took place during the time when they had the greatest appetite for risk, a period that still lingers on in teenagers of today. This behaviour is caused by a mismatch between two important regions of the brain: the limbic system, which receives hormonal messages and drives emotions (developing fully after 15 years of age) and the prefrontal cortex controlling those emotions (which does not fully develop until after the age of 25). This delayed onset of "prudence" might have been an evolutionary advantage in the past, because a greater appetite for risk is not only conducive to dangerous attitudes but also to behavioural innovations: a necessary condition for facing radical environmental changes.

We have also recently discovered that the use of recreational substances goes back a very long way. Fermented fruits were probably used in the deep past as a substitute for today's alcoholic beverages. Recent genetic studies have shown that the common ancestor we shared with chimpanzees and gorillas 10 million years ago had already developed an enzyme able to metabolise ethanol. Several of their hominid descendants were therefore able to consume inebriating substances in deep time. According to some scholars, this adaptation evolved hand in hand with a lifestyle that involved spending more time in the undergrowth, where fermented fruit abounded.

We do not know whether the Neanderthals had similar habits, but there was one important difference between them and us: we lived a little longer. Becoming fertile at around 14 years of age and being able to live to over 30 must have led to the appearance of the grandparent, a new key figure who was able to hand down accumulated knowledge and pass "wisdom" down the line.

10.3 Grandparents and Grandchildren

The advent of grandparents seems to have played a key role in spreading the innovations introduced in Eurasia among *sapiens* around 40,000 years ago. In order to improve our knowledge and translate it into successful applications, we must first master what others have done before us and then be able to absorb their achievements flexibly in order to change and adapt them to new challenges.

The potential for expanding knowledge was certainly facilitated by the presence of grandparents; it was also amplified by the larger number of young people in existence and an early familiarity with symbolic thought, a topic we will return to in greater detail later.

Not only education, but also play is very important for enabling individuals to envisage new opportunities for applying existing knowledge to previously unexplored areas. We all know how much our children love the imaginative stories we tell them and how they enjoy projecting themselves into objects (toys) and invented situations (games) that are figments of their imagination. In encouraging them, we help them develop physically but also nurture their capacity for abstraction and thus their ability to generate a world of ideas that transcends observed realities. This attribute is very useful, not only in the realm of arts, but also when it comes to translating such flights of imagination into idealised visions, tools and procedures for mastering nature and organising our social relationships.

We do not know how common it was to play in the past, but can assume that a wider pool of young people would have encouraged play and a wider pool of grandparents would have provided them more things to play with. In particular, the evolution of menopause, a trait that characterises humans alone, freed grandmothers from the care of their own children and made them available to look after the children of their daughters and relatives.

Despite the fact that Neanderthals might have possessed cognitive skills comparable to our own, they had a shorter lifespan. This is confirmed by the teeth of 70 specimens from the Krapina site in Croatia. Did a lack of grandparents and, therefore, less opportunity to learn from experience contribute to their extinction and our survival? Perhaps not, but if this factor is combined with a shorter childhood—for which there is emerging evidence—this situation could have given rise to a considerable evolutionary disadvantage, particularly after the arrival of *sapiens*. The disadvantages of Neanderthal life begin to stack up when we add to this the fact that they lived in more limited groups, while we *sapiens* were able to form more extensive bands.

10.4 Monogamy or Polygamy?

Given that we lived in large groups, one question immediately leaps to mind: what sort of sex lives did our *sapiens* ancestors lead? Opinions differ in this respect, as is often the case in the absence of any archaeological record. According to some, we would have shown a tendency to form monogamous family groups from the time of our origins, as gibbons do today. Others tend to believe that polygamy was the norm. If so, what form of polygamy—the patriarchal form exhibited by present-day gorillas, in which a hierarchical social structure is dominated by an alpha male, or a matriarchal form as seen in the bonobos, in which the group is led by a few females and the members lead varied and intense sex-lives? We do not know. What we do know is that the former is generally associated with a high sexual dimorphism, such as that characterising archaic humans.

Others believe that we may originally have had various and variable sexual relations within a group of individuals who knew each other very well and lived together for long periods of time: a kind of commune before its time. In this case, paternity would be collective and the young would be raised within the community. This hypothesis could explain the frequent difficulties in sustaining monogamous relationships in our society today and the tendency of some of us to change our sexual partners once the initial attraction has worn off and we have brought our children into the world: our lifestyles could reflect the conflict between our communal past and our monogamous present. In principle, DNA analysis of our ancestors' fossilised remains should make it possible to throw light on this issue. For example, it would enable us to differentiate between siblings and half-siblings, and more particularly, whether these were on the mother's or the father's side. However, we suspect that all the above assumptions would hold good wherever and whenever we carried out the tests. All we need to do is take a look at contemporary life to get an idea of how flexible and versatile we can be in regard to sexual matters and how varied the attitudes of different cultures and religions towards such issues can be.

10.5 Trust, Gossip and Shared Beliefs

Leaving sex aside, how did *sapiens* organise their daily lives within increasingly large groups of individuals? We are well aware that when a certain number of us gather together, we can no longer rely on relationships of individual knowledge and trust, nor can we place much reliance on the reputation of individuals mediated by the view that other members of society have of them: we need some sort of social glue. There is also a limit to the extent that we can gossip about one another.

It is easy to imagine that the first large bands of people came to be formed as a result of an increase in population, as well as through alliances with other groups of *sapiens*. This may have happened for different reasons, for example, in order to join forces so as to coordinate tactics for hunting large animals or take on rival groups. But what was the basis for those alliances? And how long could they last? A certain affinity in terms of how the world is perceived in general could have been a good starting point, but a common interest in teaming up would have been the ideal glue for strengthening such ties and making them last.

The construction of shared beliefs is a very effective means of achieving the same general world view: like-minded people tend to stick together. Trade in goods and services is a very effective means of achieving a common interest by teaming up: people who complement one another have an interest in joining forces. Many unions between couples are based on the same two principles: a similar view of the state of the world in general, as well as a taste for possible differences. In the case of alliances between groups—over and above considerations relating to a common

Fig. 10.1 Genyornis hunting scene. *Source*: drawing by Tullio Perentin, ZOIC, Trieste

philosophy of life—prevailing patterns of behaviour and economic and political relations come into play.

These are issues we deal with every day in the contemporary world. We know that a social contract is needed for a community to work. It is normally required that it be based on cooperation, but also on the punishment of any opportunistic behaviour by those who seek to take advantage of it. However, signing up for a collective ideal and a common interest works much more effectively in keeping a community together than any fear of punishment.

We cannot be certain how many individuals it takes to change a group from a social organisation based on trust and personal ties to one also based on beliefs and shared rules. Generally, however, paleosociologists consider that the number falls somewhere in the interval between societies of approximately 50 individuals ("bands") and societies of approximately 100–150 individuals ("communities" or "clans"). The optimum number for each species seems to emerge by striking a compromise between the costs and the benefits of living together. The costs arise from ecological and social competition (feeding, resting and breeding in close proximity); the benefits generally relate to a reduced risk of being preyed upon or of being expelled from a territory, as well as the opportunity to share parental care and mutual assistance.

In the case of bands, which are typical of chimpanzees and pre-human hominins, you live in the wide circle of your family and "kinfolk". Coordination undoubtedly arises out of mutual understanding, trust and reciprocity. In forming larger groups, other systems of social coordination are needed. Evolutionary psychologists appear to have established a strong correlation in the various hominids between an increase in the volume of their frontal and parietal lobes and the size of their social groups.

10.5 Trust, Gossip and Shared Beliefs

Fig. 10.2 *Sapiens* learning to fish in an organised manner. *Source*: drawing by Tullio Perentin, ZOIC, Trieste

In order to create a sense of belonging to a community larger than that made up of our acquaintances and allies, we had to maximise our capacity for abstraction and the creation of virtual worlds. To do this, we had to develop a language capable of conveying complex information.

We shall see that this information can be organised into three distinct levels: (1) information relating to the situation observed in nature; (2) information relating to ourselves and our kinfolk within the community; (3) information referring to a plethora of imaginary worlds, capable of uniting people who do not know one another but are considered members of the same culture. Some beliefs could be particularly effective in promoting cooperation with strangers, to the point of becoming adaptive traits. Indeed, it has been recently shown, through an empirical investigation, that believing in gods that are moralistic, aware of human thoughts and actions, and punitive, could be key for the construction of large-scale societies.

Later on, we will see that cultures can differ greatly from one another according to the ways in which they are built and can, therefore, serve to unite as well as to divide. This ambivalence is, unfortunately, not without consequence; this was so in the case of our ancestors, and remains as such in the complex and multicultural society we inhabit today.

Chapter 11
The Dearly Departed of the Pleistocene

We attach great importance to death. It is celebrated with elaborate rituals. Over thousands of years, we have built monumental works and complex religious beliefs based on the assumption of another life existing beyond the purely biological. Death is perhaps the area most explored by our imagination. This is probably because the topic offers very fertile ground for our symbolic thinking.

11.1 Ceremonies

The earliest traces of funeral ceremonies were found in Australia near Willandra Lakes, 800 km west of Sydney. Aborigines still revere the remains of Mungo woman, whose burial ceremony was very elaborate. Her body was first cremated and then the remaining bones were crushed, sprinkled with powdered ochre and buried. Even today, 50,000 years later, the Aboriginal people who guard this sacred area see themselves as the direct descendants of this woman, and her remains rest in a centre open to visitors, also containing the remains of other representatives of the first human groups to arrive on the continent.

As time went by, human burials involved increasingly elaborate ceremonies. In Italy, for example, we have the Prince's burial, a site found in the Arene Candide cave near Finale Ligure, dating back approximately 24,000 years. The body was that of a young man aged 15 who stood more than 170 cm high. The title of "Prince" was given to these remains, buried at a depth of over 6 m, for the opulence of the goods that were used to honour him.

His funerary offerings included four "bâtons" made out of elk antlers, a headdress of perforated shells and deer teeth, a long flint blade held between his hands, and amulets and other objects carved from mammoth ivory. The body and all the artefacts were arranged on a bed covered in red ochre powder, as in the case of the ancient Mungo woman. The young man at Arene Candide was probably killed by *Ursus spelaeus*, an animal that could weigh up to half a tonne. These carnivores used to share the Italian caves, albeit at different times, first with the Neanderthals for hundreds of thousands of years, and then, more recently, with the *sapiens*. It is thought that the Neanderthals would confront them directly, attacking in groups, armed only with Mousterian stone tools. The Prince and his group had Gravettian weapons (sharp pointed spears made out of stone or bone), but on this occasion, the bear got the better of them. The body of the young Gravettian was carefully composed, using yellow ochre material to reconstruct the missing part of his jaw, which was removed, together with his shoulder, by a blow from the animal's paw.

Many other ritual burial sites can be found in Italy. We could mention, for example, the case of Delia, a pregnant Gravettian woman who lived 24,000 years ago, discovered in the Sant'Angelo cave near Ostuni in Apulia, with the skeleton of her foetus clearly recognisable. She wore a headdress and some shell bracelets. The grave is surrounded by horse teeth and pieces of engraved flint.

Evidence of even more elaborate funerary practices has been found in other areas of Eurasia. For example, burials from the same period involving an adult man and two children (a boy and a girl) have been found in Sunghir, Russia. These individuals were wearing ornaments made up of thousands of mammoth ivory beads, hats and belts decorated with fox teeth, and other jewellery, all covered with the ever-present red ochre powder. The amount of work needed to prepare these tombs was enormous and required the support of a large community. It has been calculated that merely decorating the two young people with the 10,000 ivory beads that were found (leaving aside all the other ornaments) would have taken three man-years of work (a skilled craftsman takes at least 40 min to make a single ivory bead).

Fig. 11.1 Early funeral rites. *Source*: Drawing by Tullio Perentin, ZOIC, Trieste

One hundred and ten elaborate burials have been discovered so far, dating from the period between 45,000 and 10,000 years ago, as well as others that are less clear-cut. Various objects of daily use were most commonly found alongside the bodies. Only a few (less than 10 or so) were particularly opulent and extraordinary. Typically, the deceased were males, however, female examples have also been found.

Finds associated with a wealth of objects belonging to the deceased or placed next to them as an offering are always a source of great satisfaction for scholars. Yet, the focus is usually more on what was found (human remains, objects, artefacts and decorations) than on those who honoured them in that way. Speculating on the people who organised those burials and the reasons that may have led them to perform those practices leads us into difficult and more risky territory.

11.2 The First Hierarchical Societies

It is nevertheless tempting to put forward some general suggestions, at least hypothetically, about the type of society that could have been involved in the more complex funeral ceremonies. Firstly, we believe that in such cases there must already have been a certain surplus of resources over and above the needs of survival: the working time necessary to produce the most sophisticated artefacts means that society must have at least supported the people who made them. Secondly, in cases in which the objects were particularly elaborate, society must have been organised hierarchically. In fact, the general rule was to organise somewhat sober burials, although archaeologists tend to focus on the few lavish ones. Furthermore, in the latter cases, someone must have had the ability and skill to perform specialist work, meaning that there was already a certain division of labour within that society, or at least a pool of natural talent to draw on. Lastly, there must also have been some symbolic reason to celebrate the loss of a loved one or person of authority by performing elaborate rituals.

As far as the departed were concerned, there may have been a belief that they would enter an afterlife where they could benefit from the objects that accompanied them and rain down benevolence on those who survived them. For those left behind—their loved ones, family and clan (those who gained the greatest comfort from the funerary ceremony itself)—the symbolic reason could have amounted to an acknowledgement of their authority, superiority and social condition. This line of thought would also confirm the existence of a society that was already organised hierarchically and, therefore, was already equipped with some form of institutional structure. This is all, of course, speculation, and much more evidence is required to confirm these suggestions.

Fig. 11.2 The burial of Sunghir, Russia, about 30,000 years ago. Source: Wikimedia, author José-Manuel Benito Álvarez

11.3 Neanderthal Burials

We know that the Neanderthals also buried their dead. Some burial sites have been found in Italy, France, the Caucasus and the Middle East. It seems that the deceased were interred accompanied by rituals, although these rituals were certainly not as elaborate as those described above for modern humans. For example, the grave of a Neanderthal child found in the Dederiyeh Cave, 400 km from Damascus, contained a very well-preserved skeleton, showing that the child was buried with his arms extended and legs bent. A rectangular limestone slab was placed on his head and a small triangular piece of flint was placed over his heart. The large amounts of pollen found at another Neanderthal burial site in the cave of Shanidar in Iraq suggest evidence of a floral offering. The remains of a Neanderthal child discovered in Uzbekistan were encircled by ibex horns.

Several Neanderthal burials have also been found in Italy. One famous example is the skull found in 1939 in the Guattari Cave (on the Lazio coast, south of Rome), dating back more than 50,000 years. According to the description of its discoverer, Alberto Carlo Blanc, the skull was encircled by stones and animal bones. He suggested that this was a ritual killing during which the foramen magnum had been deliberately enlarged to extract the brain. Most anthropologists remain sceptical about this interpretation.

Chapter 12
Brain Readers

We have a friend who is a good magician. He asks us to think of a card and then unhesitatingly picks it out of the deck. We have never worked out how he can read our minds. Beyond such tricks, we now have technologies that can be used to try to understand the workings of our brains, even reading our fears, hopes and illusions. But can we enter into the heads of our distant ancestors from deep time? Incredibly enough, we can not only read their brains, but also their minds. To begin with, we need to establish the difference between the two concepts.

Leaving aside the centuries-old philosophical debate about this subject, some use the language of computer science and consider the mind to be the software that runs on the hardware of the brain. The concept of mind can then be extended beyond the boundaries of the brain to include the use of external objects. Hence, our cognitive abilities depend on an integrated system made up of brain, culture and the environment: brain and culture connected through our bodies; culture and environment connected through the objects we use. We will see that these objects can nurture, but also curb, our mental abilities. Specifically, culture, environment and associated objects can lead to changes that can be passed down even if they are not directly related to the human genome (through epigenetic mechanisms, processes that produce new traits—anatomical or behavioural—without DNA mutations).

12.1 Inside the "Grey Box"

Modern neuroscience describes the manifestations of the brain in terms of biological elements such as neurons (the special nerve cells found in the brain) and synapses (the connections between cells), both of which are components of very complex circuits. According to the most recent estimates, the brain contains 100 billion neurons, each connected to thousands of other cells through hundreds of trillions of synapses distributed throughout the cortex. They carry nerve impulses to neurons and other parts of the body, such as muscle fibres. It has also been found that the brain operates using specialised modules that are interconnected. This division of labour, typical of a complex system, increases the efficiency of information processing.

In medical applications, the brain's functions are analysed using neuroradiological methods that enable its activity to be viewed in vivo. One of the first such techniques, introduced during the 1960s, was brain scintigraphy. This is based on the use of radioactive tracers, which concentrate in different regions of the brain according to blood flow and other parameters that describe the organ's metabolic and biochemical activity. This method was used to confirm correlations between language and the morphology of the brain's left hemisphere suggested by Paul Broca in the nineteenth century after he had studied a patient with an extreme inability to pronounce words. In the past, abnormal conditions were the main source of information on the workings of the brain.

In the 1970s, neuroradiology began to make use of more sophisticated methods, such as positron emission tomography (PET). This technique involves placing a radioactive tracer emitting positrons (anti-electrons) in the brain. It reveals the way different parts of the brain work, following specific chemical processes that regulate the various brain functions. Other brain imaging methods are based on CT scans, using x-rays and magnetic resonance; both of them allow us to study the anatomy of the brain structure in detail.

PET and other techniques allow us to "see" which parts of the brain are activated when we experience different emotions. We can also identify the problems that affect our health, our cognitive and memory skills, and our mental balance. We can even "photograph" disorders such as schizophrenia and the various agnosias (in other words, the inability to recognise objects, people, shapes, smells, and so on). Recently, these methods have been used to communicate with patients in a vegetative state.

Psychology itself is now becoming part of neuroscience, based on simulations and computer models that connect it to the other "hard" sciences. Even genomics is becoming part of neuroscience: we can now identify the genes responsible for brain development, and we know that these account for nearly 59 % of the approximately 20,000 genes in our genome. This has led to the development of a new interdisciplinary area known as "molecular psychology", in which data obtained from neurology, which tell us about the brain structures and functions we mentioned earlier, are cross-matched with data obtained from analysing the human genome.

12.2 Mind and Brain in Deep Time

While neuroscience mainly applies to present-day human medicine, it is also applicable, to some extent, to the study of our ancestors' brains in deep time. This statement may seem surprising. How can we obtain structural and functional information on a brain, which is one of the more perishable organs and certainly not

12.2 Mind and Brain in Deep Time

preserved in the geological record, long since turned back into dust? In fact, there is a way of achieving this.

Information can be obtained on an important part of the brain: the cerebral cortex, which is the wrinkled external part in contact with the skull that changes in thickness throughout life and plays a central role in the cognitive functions governing consciousness, language and memory. Important information on the structure of the brain (strictly speaking, we should call it the encephalon) can be obtained by analysing the inner surface of fossilised skulls. Advanced CT scanning systems based on synchrotron light can now illuminate (literally) aspects that were previously inaccessible.

We have already talked about these micro-CT techniques with regard to the analysis of hominin teeth. If we now apply them to fossilised skulls, they are able to provide us with additional information and, more importantly, they can do so non-destructively. In the past, the risk of damage to these precious specimens greatly slowed down studies or prevented them from being attempted using the methods then available—such as casting the inside of the skulls with moulding materials. Nowadays, the problem no longer exists, and high-resolution 3-D images of the skull can be generated and virtually sectioned. Hidden details can be extracted to make quantitative analyses and comparisons.

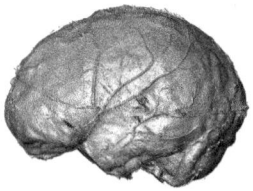

Fig. 12.1 (**a**) and (**b**) Reconstruction of hominid brain using x-ray microtomography. *Source*: Reconstruction by F. Bernardini et al. (ICTP, Trieste)

In practice, even though the brain is long gone, we are still able to obtain a perfect and highly detailed virtual endocranium onto which a mould of the cerebral cortex is imprinted. It is ultimately possible to see many details that were impossible to obtain using traditional methods. We can see, for example, the vascular and connective elements of the brain, or at least those that are imprinted on the inner surface of the skull through the "dura mater", which is the outermost part of the meninges (the membranes covering the brain). Since we now know a great deal about the functions performed by the different parts of the brain, the internal details of the skull cap provide us with very detailed information about "how" and "what" the skull's owner was able to process mentally. We can even hazard a guess as to whether he (or she) had a flair for art or music, if he was a good speaker, if he had some other natural talent, and how much he was able to learn.

But we must not get too carried away. Many internal details cannot be explored—and even though some brain functions, to some extent, can be associated with a specific area of the brain, we cannot obtain accurate information on the all-important connections between them in the absence of an observable functioning brain.

However, it is hard to overestimate the importance of these new technologies, especially if the results can be linked to those emerging from other related disciplines such as genetics. As we saw earlier, we can sequence DNA in samples of human remains that date back hundreds of thousands of years. The entire genome has now been mapped for various individuals of the *H. sapiens* and *H. neanderthalensis* species, as well as for the Denisovan individual belonging to the new human form recently discovered in Siberia, though we do not yet have a skull for the latter species.

A new line of research involves the collaboration between neuroradiologists and archaeologists. It allows us to identify the neural circuits activated in a present-day knapper, when he is set to fabricate a stone tool with the same techniques used by ancient humans. One can see, for example, that more and different parts of the brain are activated in accordance with the degree of complexity of the tools to be made.

12.3 Building Up the Human Mind

The evolution of the human brain has always been a topic of crucial importance in the study of human origins. After all, it is used to define what it means to be first "humans" and then "modern humans". Its growth from 350–400 ml in the first hominins to 1350–1500 ml in recent species of *Homo* has been widely discussed in the previous one and a half centuries.

A new interdisciplinary approach, which we could call "molecular paleo-psychology", allows us to cross-match data on the external brain structure of various human species with data on the genetic structures responsible for their various cognitive and behavioural functions, and then cross-match both of these with archaeological evidence. The latter could include the level of complexity of the stone industry, the use of specific materials and personal ornaments, the presence of artistic endeavours such as wall painting and figurine crafting, the creation of musical instruments and even the evidence of elaborate ceremonies for burying the dead. What can this work of connecting and cross-matching data tell us? We will start by looking at cognitive development.

The cognitive development of our ancestors can be inferred from their ability to carry out complicated operations, particularly those requiring a very sophisticated type of communication. These undoubtedly include the ability to improve long-distance navigation. It is hard to believe that this would have been possible without a degree of symbolic thought. For example, getting to Australia required the ability to cross long stretches of open sea and to envisage the presence of land beyond the horizon. The descendants of the African *sapiens* managed to do it, exhibiting an ability to imagine (and navigate) that was lacking in *H. erectus*, who lived on the Indonesian islands for over 1 million years without ever crossing the seas that separated them from Sahul. The Australian Aborigines still speak of "dreamtime" when referring to their origins.

Some argue that molecular palaeopsychology may be able to help us understand whether the evolution of the modern *H. sapiens* mind took place in the wake of sudden genetic changes that introduced new brain structures and connections giving rise to new social and communicative forms of behaviour, or whether it was due solely to cultural, demographic and economic mechanisms that remained dormant for many millennia. However, the evolutionary path of our mind does not seem linear. Complex feedback mechanisms might have come into play. Some of them were certainly of a genetic and psychological nature; others stemmed from the social and economic environment that we managed to establish in order to increase our survival chances. It is only by studying their interaction, we argue, that we can make some sense of the virtuous cycle that has led our mind to be what it is.

The latest methods of studying the brain can be extended to investigations of pre-human hominins. Recent CT scan studies using synchrotron light on the skull of *Au. sediba* show that his brain, which was similar in size to those of present-day great apes, was characterised by slightly asymmetrical frontal lobes similar to those of humans. This suggests that the evolutionary line preceding *Homo* had probably begun a process of neuronal reorganisation even before the brain volume expanded and that this may have involved an increase in connections in the region associated with language and social behaviour in humans. The separation between humans and the previous hominins, who perhaps already knew how to make stone tools and had some cognitive abilities, therefore becomes increasingly uncertain.

12.4 Thinking About Thought

Some have associated the size of the frontal and temporal lobes—crucial for cognitive functions—with the size of social groups and the "levels of intentionality" we can infer with regard to the thinking capacity of various hominins (and even the apes). Intentionality is a term that refers to the ability to have opinions about oneself and other people within a certain social context. This feature is closely linked to the capacity for imagination and is often associated with statements such as "I believe", "I think", "I want" or "I suppose". It also concerns what other individuals believe, think, want or suppose about themselves and others—and applies to the psychological development of our children as they grow, meaning that they possess a "theory of mind". This capacity is hard-wired into our cortex through the "mirror neurons", which allow us to perceive somebody else's feelings (often subconsciously) as if they were our own and make us identify with what we observe.

Going back to our mind, at the first level of intentionality, reflected, for example, by a macaque and a *sapiens* child under 3 years of age, we only have a certain self-awareness and are unable to conceive of a world other than the one we observe. At the second level, typical of australopithecines (and maybe even a chimpanzee) and a *sapiens* child aged from 3 to 6, we begin to have some awareness of what other people think. At the third level, typical of *H. erectus* and a *sapiens* child from 6 to 8, we make a number of assumptions about what others think of us or about other people. At the fourth level, supposedly typical, for example, of *H. heidelbergensis* and a *sapiens* child from 8 to 11, we can also think about the thoughts described above. At the fifth level, typical of *sapiens* (and perhaps even Neanderthals), our desire to make people believe that all the above is true may come into play. In short, it is at this last level of intentionality, and possibly at levels beyond it, that we see the advent of collective values and ideals. This is fertile ground for the emergence of structured ideologies and religions.

For example, organised religions can be said to belong to (at least) the fifth level of intentionality, if some individuals set out to persuade other individuals of the idea that if they appeal to a higher Being, they can induce it to act in the manner they want (and pray for). In the case of religions that are also "revealed", we ascend to another level. By extrapolation, when we authors set out to persuade you readers that religions operate at the previous level, we go up by one level of intentionality (to the seventh level). Readers might apply this reasoning to the case of ideologies, all the way down to TV advertising.

Even scientific thought operates at very high levels of intentionality, but differs from religious thinking, as it needs to verify everything the human mind is able to conceive.

If the above is borne out by further studies on brain faculties associated with different types of hominin, it will become more difficult to establish a certain date marking the start of our divergence based on brain capacity. At least, though, we will be able to set a more reliable date for the advent of our capacity for abstraction, which led to the creation and coordination of increasingly complex societies: that date marks the appearance of modern humans.

Chapter 13
All Power to Imagination

If we imagine a *sapiens* looking at an animal and drawing it on a wall, we say he is making use of symbolic thought. That drawing is the outcome of his interpretation of the observed reality: an abstraction that turns a living being into a graphic representation. Two realities thus exist: the observed and the represented. If the process had ended there, nothing would have happened. The artist would have been ignored or even scorned by his contemporaries and his descendants would have seen strange scribbles on a wall.

But if the observer of the drawing had also been endowed with a similar mental capacity and an appropriate key for translating what is observed into what is pictured, then he (or she) would have immediately seen two realities: the three-dimensional reality of his surroundings (from which the artist drew inspiration) and the two-dimensional reality portrayed on the wall. At a certain point, it might have dawned on the observer that he could also have a go at representing the reality around him. Even though he might not necessarily have been a good draughtsman, he could always be a sculptor and make a statue that depicted something he had in mind and was currently obsessed about, such as an erotic representation of the female body. Or he could be a musician and compose a sound sequence that conveyed a state of mind or a vision of the world. If he were a poet, he could imagine that someone had named things by singing about them and thus brought them into existence (this view is illustrated by the mysticism of the Australian Aborigines, for example). He could even be a skilled craftsman and busy himself with innovatively combining various materials (stone, wood, bone and skin) to build a variety of products useful for his daily life. In practice, each would have been able to generate his own "virtual space" in which he could imagine what he was about to do (or say).

As with the sexual reproduction of most living beings, the initial contribution of at least two individuals is required to ensure that symbolic thought is replicated and expanded throughout all possible fields of application: in this case, the person who "represents" and the person who "recognises" this representation of thought. All it takes to be able to convey this composite thought to other individuals is a common language. The process will obviously be easier and faster if the individuals in question are able to glean an idea of what their interlocutors are thinking, as emphasised when we discussed the different levels of intentionality that characterise social interactions.

The ability to spread and articulate symbolic thought will also increase if it is possible to reduce the barriers between individuals in space and time. This applies if the individual who represents an idea and the individual who recognises and interprets it do not necessarily meet physically or live at the same time as one another. This might be why we sometimes associate the beginning of "history" in the strictest sense with the invention of writing, which allows ideas to be conveyed through time and space. The current contactless expansion of ideas by way of information technologies is happening right before our eyes. Everything began, of course, with the invention of spoken language. Without wishing to underestimate its importance, we are bound to acknowledge that a spoken language is a volatile and ephemeral system of communication. We can try out other methods for conveying an idea: drawing it on a wall, passing it down through the generations as a legend, summarising it in a song, tattooing it on your body, or including it in a ritual.

As a specific example of the complexity of the interactions that come into play in evolutionary terms, writing and reading (in a broad sense) not only broadcast new ideas, they also promote significant feedback effects on brain functions, an all-important mechanism that is often overlooked. Psychology provides us with important information on this subject. For example, it has recently been shown that "reading" what is "represented" by other people in all possible forms (written or otherwise) helps the brain capacity to expand in many directions. Firstly, it increases empathy, making us see the world through the eyes of others. Secondly, it gives the brain a great workout, in the same way that playing chess or solving a crossword puzzle does. The faculty of expression and visualisation in children (confirmed by means of magnetic resonance techniques) is enhanced as well. Lastly, it helps to seduce, because it promotes intelligence (analytical and emotional) and our understanding of others. To sum up, symbolic thinking not only helps convey ideas; it can also have important repercussions on the "machine" that generates the thought itself, ultimately improving it.

Unfortunately, all these beneficial effects seem to apply less to the latest generation of digital technologies: in the absence of adequate interpretative skills and imaginative stimuli, an excess of unchecked information may indeed reduce the beneficial feedbacks described above.

13.1 New Realities

If one believes that "imagination is more important than knowledge", it should be possible to generate realities capable of bringing people together around a shared idea.

According to this viewpoint, we should be able to start with miths and religions and end up building ideal societies. The only requirement for initiating the process would be the development of a convincing symbolic representation and its reproduction, which would be "shared" by the observer (meaning that he would believe in it or be compelled to do so, due to fear or reasons of convenience). By generating principles that can be used as a basis for social behaviour, we can form what we now know as "institutions", meaning not merely the bodies that go under this name but the set of rules that a society decides to respect and enforce. Although different in kind, believes and rules of behavior are often identified with one another, even though the former aim at describing what is considered to be "true" and the latter what is suggested to be "right".

The construction of myths and religions began by borrowing images from nature (the Sun, the Earth and some animals), perhaps combining them with human characteristics. Some dared to suggest that divinities shared the appearance and character of humans, with all their strengths and weaknesses. Alternatively, man was supposed to be made in God's image. Yet, some refused to accept this bond between images and beliefs and forbade representations of the deity in any form, since it was seen as being exclusively part of the spiritual realm. The matter is not settled, however, and we are surrounded by an overwhelming representation of deities. We need symbols to convey ideas.

When it comes to the creation of institutions (in the sense of codes of conduct), the question becomes even more complicated. At some point, we must resort to constructing symbols that are capable of defining a relational structure that is acceptable to the people (willingly or by force). For example, we shall see that it is possible to greatly increase the availability of goods and services within a society through the division of labour. In order to distribute those goods and services to the members of that society, we have to imagine a set of rules that may be based, for example, on a kind of hierarchy (for which we have to find some excuse); alternatively, we could imagine societies based on equality. In the former case, a coat-of-arms or a flag referring to it could be considered an appropriate image to represent that idea, while working tools could serve the same purpose in the latter case.

Once the institutions have been imagined, they could be "acknowledged", and thus give rise to kingdoms and homelands, as well as legal, economic and political systems based on a set of ideals and shared principles. We could invent institutions like the States or the United Nations and ethical systems of reference such as "human rights". This could lead us to bomb, perhaps under the auspices of the United Nations, a population that is alive and thriving, just because that State does not respect "human rights". Thousands of people could be driven to sacrifice themselves for a religious belief and massacre the same number for a principle, such as freedom, equality or fraternity.

Structures that have never previously existed, such as limited liability companies, could be invented and given legal status, so that only their managers and not all their owners are legally responsible for their actions. Large industrial companies (made up of men, machines, skills and products) could be separated from their physical and organisational structures and identified only by their symbols (logos).

This action would make all the material and human components of a company interchangeable and would allow it to survive (or fail) merely to the extent of human confidence in that logo. Value could be attached to certain symbolic conventional objects (turning them into money, for example) to ensure that everyone exchanges them at that particular value. Wealth (in virtual form) could even be generated through electronic transactions without any exchange of goods or services taking place.

In short, there is unlimited scope for proposing possible worlds if we accept that these worlds can be located only in our minds. By doing so, we make them real, but on one condition: that they are first "represented" with an image and then ultimately "acknowledged" and "shared" by a community as a whole, thus helping to form its cultural frame of reference. Despite living in a "natural" world made up of four dimensions (if we include time), we have ended up adding a fifth dimension to our lives: one generated by our minds.

It is striking and even a little daring to note that in the Creation of Adam painted by Michelangelo in the Vatican's Sistine Chapel, the image of God and the surrounding angels are painted within the shape of a sagittal section of the human brain. In particular, an expert might discern the cingulate gyrus in God's left arm, the vertebral artery in a swirl of fabric, and the pons in an angel's back; even the spinal cord could be represented by the legs of one angel and the structure of the pituitary gland by the cleft foot of another.

The current interpretation is that Michelangelo, who was also an expert on human anatomy, set out to paint God in the act of conferring intellect on man. Yet, the fact that God, and not Adam, was painted within an allegorical image of the human brain suggests a topsy-turvy interpretation of the meaning of that fresco, in other words, that the intellect of man (in this case, that of Michelangelo himself) was responsible for generating the image of God.

We leave readers to make what they like of these suggestions, and proceed to the next important question: "When and how did these new forms of behaviour associated with symbolic thought first arise"?

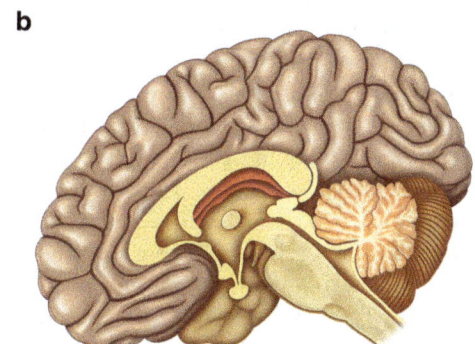

Fig. 13.1 (**a**) The Creation of Adam, painted by Michelangelo, in the Vatican's Sistine Chapel. *Source*: Wikimedia commons, Author Jörg Bittner Unna
(**b**) Brain sagittal section. *Source:* Shutterstock.com, Copyright Alexilusmedical

13.2 Inside the Black Box of Symbolic Thought

In order to apply symbolic thinking in different fields, at least two conditions must be met. The first is that there must be an opportunity to convey thought through a shared language. It is important for this language to be an open system that allows different meanings and interpretations so that it can be sourced and spread with everyone's help. If communication is not carried out through language but by means of a code, for example—such as when we communicate with a computer through a keyboard—this amplification would not take place. The second condition is that individuals should differ from one another in terms of aptitudes and abilities. This allows them to better articulate their own thoughts when they access the thoughts of others. This idea can be represented as a bâton that changes shape when passed from hand to hand in a relay race. The multiplying effect of this process will be high when applied to a growing number of individuals. In the end, it will generate a very complex reality that is the outcome of all our different representations.

We have also seen that it is possible to represent something that exists only in your own mind: for example, a figure with a human body and a lion's head (such as the ivory figurine found in a cave in Stadel, Germany, dating back 32,000 years). This paves the way for representations reflecting different degrees of realism, until we reach the point at which we conceive realities that exist solely in our imagination. It is ultimately possible to create events, circumstances and procedures that are based only on this imagined reality. One such example is burial rites based on the idea of life after death. It is also possible to tell stories about things that have never happened, such as events in myths or legends.

Once this point has been reached, it is then necessary to be able to discriminate between the representation of something we have actually observed and the representation of something that we may have imagined. How do we do this? Answering this question is not a trivial task, and would lead us to stray far from the topic if we attempted to do so here. Focussing on the theme that concerns us, which is that of human evolution, we can safely say that we often fail in this regard; we assume things to be true when, in fact, they are only flights of fancy. Usually, this happens because believing brings us tangible benefits, increasing our material and spiritual well-being. In particular, we will see in the next chapter that our prosperity increases when we believe in something that facilitates the proliferation of trade in goods and services. Our spiritual well-being increases, on the other hand, when we are able to rely on the all-powerful guidance of higher Entities, an important factor in the construction of different religions. In this case, we can see our spiritual well-being as a substitute for our mental well-being.

13.3 Contemporary Examples

Looking at the present-day, the best example of how symbolic thought is currently expressed can be seen in the behaviour we refer to as "creative", which is typical of many professions such as engineering, design and advertising. It takes the form of inventing new products and establishing new needs to enrich our lives and our social relationships. It focuses on innovation, design, fashion and status symbols. The aim of these professions is to satisfy the image that individuals wish to project of themselves when they act as consumers, and possibly create new ones. Today, we are so fascinated by these imaginary representations of ourselves, and the prospect of a completely unreal lifestyle, that we allow advertising to invade our homes without the slightest concern for the sense of dissatisfaction we then feel when it is not possible to live up to the images we are shown and required to identify with. We also often make the mistake of believing that the price of a commodity always represents its true value, and that if something is more expensive, it must surely be of a better quality, as if the underlying condition (the ideal of perfect competition) would hold in the real world of business. This does not happen to capuchin monkeys, who, according to a recent experiment, are very well able to discern the quality of the products offered to them and pay a fair price when they are trained to understand the value of the currency they can use to buy them.

We represent ourselves creatively (or think we do so) when we want to socialise. We do this through the clothes we wear, the objects we use, our means of transport, our membership in some particular group (my social class, my football team or my country), etc. We can even create an avatar of ourselves by selecting only certain information about ourselves and sharing it on Facebook with hundreds of virtual friends. We end up living in a fantasy world. We have slipped through the looking glass with Alice. A substantial part of the society in which we live is manifested in this world of representation. Our institutions (in other words, the codes of conduct we follow) exist in this world. Those who are inspired by different religions and ideologies also live in this world.

Yet, we would be making a big mistake if we believed that this behaviour is anything new. It goes back to when we invented the first body decorations, painting our skin and adorning ourselves with jewellery, tattoos and garments with aesthetic value. It goes back to when we began to paint, to play music, to create the first rules of relationships and coexistence; when we began to hand down myths and legends to create a shared memory—and when we created the first "virtual" realities corresponding to our need for transcendence, spirituality and socialisation.

Fig. 13.2 Woman from Karo tribe with her child in Kolcho, Ethiopia, 12 August 2014. *Source:* Shutterstock.com, Copyright Luisa Puccini

13.4 Possible Explanations

We still cannot agree over the reasons underlying the behavioural patterns characterising modern humans.

Some have argued that when we were beset by great environmental adversities, another evolutionary tool was required to operate alongside the survival strategy based on natural selection. This other tool is known as sexual selection (also studied by Darwin). The two selective mechanisms are actually different. Natural selection depends on successful adaptation by both genders (male and female) to the environment, while sexual selection occurs during competition between individuals of the same gender for mating purposes. It is designed to give the greatest chances of reproducing to the most creative individuals of the same sex: for example, those best able to represent themselves through ornaments, body paintings and objects designed to attract the attention of the opposite sex. You could even say that body art preceded or coincided with the first forms of artistic expression. These changes effectively made it possible to seduce with the mind. This trait was then genetically passed on to subsequent generations and would find applications in many other fields.

A second explanation refers to the ability to create a complex language suitable for interaction in a growing and structured social organisation. By the time very large groups of individuals had formed, a system of communication capable only of describing the outside world (such as the alarm signals common to many animals) or a language that allowed us to exchange information about ourselves and our acquaintances (which can only work in a society made up of a few individuals) would not have been enough to build relationships based on trust. A need had arisen to invent a language capable of conveying that which does not exist in order to consolidate and organise a multitude of individuals: an ideal construct.

What benefit would this approach have brought? The answer is quite simple. If you can influence social behaviour by telling stories, you can change social behaviour by changing the stories you tell. As mentioned before, biological evolution may have been progressing too slowly, and this time, we might have used cultural evolution to speed up the acquisition of new forms of behaviour, something that is only possible through symbolic thought.

The two approaches described above—one based on sexual selection and the other based on cultural evolution—are not mutually incompatible: developing a complex language also has a very persuasive effect on the mind and helps us seduce prospective partners. By now, we had a highly developed cerebral cortex that had been honed to a peak by natural selection over the previous 2 million years and its potential could be unleashed in the fight for survival brought about by the dramatically fluctuating environmental conditions. If nature was changing too quickly, we could do so too, cooperating within larger and larger groups.

The evolutionary strategy of developing one structure for one function and then using it for a different purpose can be compared to the innovation of feathers in some dinosaurs. These were originally used to improve thermal insulation but then proved essential to facilitate flight in the birds that were their direct descendants. Similarly, our brain did not grow with the aim of developing symbolic thought, but at a certain point, we were able to achieve this as a by-product. We may not have feathers, but we can still take flight through our imaginations.

The importance of symbolic thought does not only concern artistic or economic representations. It also concerns the advancement of knowledge in science. By allowing us to conceive worlds we cannot see, we are freed from the shackles of false information provided by the world we do see. When this information is filtered through our limited and imperfect senses, it is easy to believe that the sun revolves around the earth, that the earth is flat, and that the stars are lights hanging from the sky. Even scientists and philosophers are sometimes fooled by perceptions: it took almost 2000 years (from Aristotle to Galileo) to understand that the force imparted to a body is not proportional to the speed acquired by that body but to its acceleration.

Our senses also prevent us from appreciating reality outside a narrow range of colours, sounds, smells, tastes and tactile stimuli. If we are to understand the laws of nature, we must use our imaginations to go beyond the world of sensations and perceptions created by our brains based on the interpretation of information gathered by our senses. By doing this, we can even "see" invisible components of matter such as atoms and quarks.

Because new information and communication technologies give us opportunities to reach a growing number of individuals, we can exponentially accelerate our world views so that everyone can help them to build up and expand over time. Symbolic thought becomes the most precious "common heritage" of mankind.

13.5 Symbolic Thought and Cultures

We could argue that when certain forms of behaviour take root in a group of individuals and are handed down from generation to generation, a culture is also formed: something that is partly stable and partly continues to evolve with the passage of time as new ideas become established in society. These ideas appear to operate in a manner similar to gene mutations in biological evolution, but—unlike mutations, which are random—the new ideas often come about for very specific reasons and are adopted only if they "work" for the purposes of survival. According to this definition of culture, some forms of behaviour—for example, the use of fire or the fashioning of the first stone tools—could take root and be handed down from generation to generation simply due to their practical usefulness. This approach is typical of cultures that came before the formation of symbolic thought and did not need specific changes because they were well equipped to guarantee survival in the absence of major environmental challenges.

There seems to be a fundamental difference between symbolic thought and culture. While the former is now of universal significance to modern humans and goes hand in hand with our emerging capacity for abstract thought, the latter is defined within a limited time and space. In other words, symbolic thought is a cognitive process while culture is often (though not always) a product of this process. The former needs to emerge and spread freely; the latter needs to take root. The former consists of a flow of ideas that multiply; the latter consists of a stock of knowledge to be shared. If one insists in overlapping the two concepts, he must be aware that he is describing both the statics and the dynamics of a complex evolutionary process.

Different cultures (in the sense described above) began to proliferate in all the areas populated by *sapiens* halfway through the last ice age, with the greatest hoard of archaeological evidence appearing in Eurasia roughly 45,000 years ago. The various cultures that succeeded one another on a scale of increasing technological and artistic dexterity were called: Châtelperronian, Uluzzian, Aurignacian, Gravettian, Solutrean and Magdalenian. The latest cultures were marked by realistic representations of animals of the glacial period, erotic ivory figurines, and flutes made out of bone and other materials, as well as tools for fishing, hunting and making clothing.

This does not, however, mean that all other cultures, even the most ancient in which symbolic thought appeared (in Africa, Asia and Oceania), can be compared with the European experience to create a hierarchy of lower and higher cultures. Nor does it mean that more archaic cultures in which symbolic thought was in its infancy or even non-existent are not interesting. In the latter case, the culture was expressed only through information and technology that were replicated with minimal improvements, as in the case of Acheulean culture/technology, which lasted one million years. In any case, we were certainly not the only species to have the sort of culture we have just defined. We also apply the term culture, from this perspective, to many primates and even elephants, which have recently been shown to behave highly disruptively when orphans are removed from their slaughtered matriarchal families and can no longer learn social codes of conduct and information relating to survival.

All we know is that our symbolic thought originated, approximately 100,000 years ago, somewhere in the extensive area represented by Africa and the Middle East. As it evolved from an individual to a collective form, it progressively developed and extended into an increasingly complex and seemingly unending series of representations. Initially, we tend to admire them. But after a while, we may feel uneasy about them. They seem to have gone over the top, even spiralling out of control.

13.6 Excesses of Representation

What is the explanation for the hundreds of thousands of monuments, statues, portraits and objects worked and embellished using increasingly refined techniques that have been handed down to us by the various known civilisations, of which only a small fraction is exhibited in museums? Why build 8000 life-sized terracotta warriors, each different from the other in the tiniest detail (even down to the shape of their ears), to represent the Chinese Imperial Army in the third century BC? What are the origins of this desire to celebrate excellence and abundance beyond all limits of common sense?

One possible general answer to all these questions could be the need to form a collective identity that draws on the fierce pride of belonging to a community and coordinating individual behaviour within an increasingly extensive and complex society, thus building its cultural heritage, to be first disseminated and then, if necessary, reformed, recycled or destroyed.

Such excesses of representation are not effectively limited to art forms; they also extend to economic attitudes. We need only think of our actions when we collect property, currency and securities in amounts that enormously exceed all our possible current and future needs. In this way, we end up with wealth to pass on to our descendants, who have sometimes done little to deserve and do much to squander it. Another example is the creation of relationships, places and spaces that amount to huge infrastructures designed to meet our individual and collective needs for interaction and exchange, such as the large metropolitan cities.

Our creativity has led us to trade financial products with an estimated value that is far higher (some say 10 times higher) than that of the real estate and production capacity that they are meant to represent in the form of a security or other financial instrument. This over-representation has now become a problem that we refer to as the "financial crisis", which we do not know how to resolve. Over the past 30 years, we have allowed people to circulate securities that correspond to a particular type of "real" wealth, as well as securities that merely represent other securities, and this has happened many times. This chain of events has destroyed any meaningful connection between the real wealth and the financial wealth that we were so keen to create in the world of our imagination over the last few centuries. Those who hold financial wealth and trade in it still believe they are circulating the ownership of a company, a property or a credit. If they do not, they think they can fool somebody else into believing it. But this is not always the case. We will not have any problems as long as we all live under this illusion, though one day, our chickens might come home to roost. In this case, if the above calculations are correct, nine traders out of ten will have to wipe the virtual wealth from their portfolios. Due to the way our economic system is built, and the way in which has become totally inter-dependent, we will not be able to shrug this off as "their business". Our over-representation must be curbed, but unfortunately, we have not yet discovered a way to do this.

13.7 A Recap

Going back to the roots of this behaviour, we can conclude that human history unfolded due to four specific evolutionary conditions. Three of them were "necessary" but not "sufficient", using the terminology of philosophical logic. They were: (1) the development of an upright posture, which freed the upper limbs from the demands of locomotion and improved the use and handling of tools through significant anatomical innovations; (2) the growth of a brain that was "large and powerful", both in terms of volume and in terms of neurons and synapses engaged; and (3) the establishment of a vocal structure that was capable of delivering a language to convey information about the observed reality, about ourselves and about the people and things we wanted to talk about.

Inventing a language, though, was not an easy task. It required brains. To put it simply, we had to associate a sound with what we wanted to talk about and share it with others according to certain predefined rules. This already implies a certain degree of abstraction: we had to invent a "world of words" and connect through it. This is probably when a new dimension to the reality in which we live began to form, but it would take a long time before we would be able to fully exploit its potential.

In a society divided into small groups, and in the absence of any major environmental challenges, this was all we needed. However, if we had remained at this stage, we would be barely distinguishable from many other illustrious exponents of the animal world. As *Homo*, we would have had some extra leverage in terms of our expressive language and our manual dexterity, but it would have been difficult to envisage a future in which we became "masters of the Earth". At some point, however, conditions arose (environmental and social) that facilitated the development of new cognitive abilities.

The necessary and sufficient condition that made us modern humans so different from all other species is the full development of symbolic thought. This ability can be associated with the growth of certain areas of the brain and the feedback mechanisms that arise during the interaction between our brains, our culture and the environment. This capacity allowed us to create worlds that transcend observed reality. This is handed down and multiplied through increasingly complex forms of language, including spoken language, written language, music, dance, figurative and abstract arts, mathematics and computer languages.

To sum up, the full extent of symbolic thought allowed us to assemble a combination of traits that eventually seems to have given us the edge:

(i) a long period of apprenticeship for new generations, which allowed more time to learn old and new tricks, and thus enhance the creation and accumulation of knowledge;
(ii) an advanced cognitive capacity, which can multiply and accelerate when passed over from mind to mind, especially when it is meant to last, and is thus made independent from time and space;
(iii) the ability to form large groups in which to cooperate, but also the ability to compete with other groups for access to limited resources, both abilities moulded to be variable and flexible, in order to adapt to circumstances.

Nowadays, at the end of this long process, we convey our thoughts in forms that we do not share with any other living species, even though we may have partially shared them at one time with other humans. This collective cognitive process has probably enabled *sapiens* to continue surviving and prospering while living in increasingly extensive and complex societies. It remains to be seen how long we can continue to do so without changing the rules of engagement in our endless struggle against nature and people who are not "our own".

Chapter 14
Primordial Economy

Who would have thought that Leonardo's *Mona Lisa* or Michelangelo's *David* were the final outcome of a form of behaviour that might date back to the beginning of the last ice age? And could the emergence of artistic behaviour actually have coincided with the appearance of economic behaviour? Incredibly enough, this seems to have been the case. But what is meant by economic behaviour? The most popular and general answer usually refers to the rational pursuit of self-interest. We have touched on some examples in the previous chapter, and now, it is time to delve into the specifics. A more precise answer depends on the time scale considered.

14.1 *Homo Economicus*

If we refer to the short-term, in which we can satisfy our needs using only readily available resources (by hunting, gathering the fruits of the earth and making everything we want to produce), behaviour is defined as economic when we seek to derive the highest individual satisfaction from the most effective use of all those resources. Satisfaction obviously depends on the needs and tastes of each individual, but above all, on their ability to give something in exchange for what they desire. Since needs are always considered to outweigh the potential for satisfying them, economic behaviour boils down to a simple choice between the goods and services available, limited by one's ability to pay for them. This quintessentially individual viewpoint sheds light on the role of people (economic) as striving to increase their well-being, making a virtue of necessity (in other words, being content with what they can afford).

If we refer to the longer-term, which seems more appropriate to our topic, the available resources obviously vary: for example, they may increase due to our efforts to obtain them or decrease as a result of wars and famines. In this case, the form of behaviour described above does not make sense. The purpose of economic behaviour would be exactly the opposite: to increase the quantity and quality of goods and services that satisfy our needs. This would mean we could rescue ourselves from the constraints imposed by nature through our work and ingenuity. We shall see that this result is much easier to obtain if we do not have to rely on ourselves alone but are part of a society that is capable of enhancing individual talents. This approach creates a collective identity: we become part of a social body.

The division of labour, described so ably by Adam Smith and David Ricardo, demonstrates how easy it is to satisfy our needs by working less and producing more: all it takes is for everyone to specialise in what they do best, pouring this knowledge into increasingly detailed tasks and creating more and more specialised products. The result will be a boom in goods and services available to society as a whole. Then, all people have to do is exchange these products in accordance with predetermined rules, one of the main applications of our capacity for abstraction. The results will be a more or less egalitarian society, depending on the prevailing economic and political system.

Nowadays, the division of labour and technological progress have dramatically increased the availability of goods and services, but also led to extremely complicated products that a single individual could never begin to know how to make if working alone. We are living in a world of continuous trade. But when did this custom begin, and what prompted it?

14.2 The Origins

We have seen that, during their first 100,000 years of history, *sapiens* went on developing their lithic technology very slowly, and symbolic representations were scarce. At that time, our ancestors were *sapiens* like us in their anatomy alone; definitely not in their mental abilities. They had a culture, perhaps with the first hints of symbolic thought, but probably did not display any economic behaviour when relating to one another, unless we define this simply as a one-man drive to make ends meet in bad times. What was the crucial tipping point? We previously asked ourselves how symbolic behaviour led to the various cultures. Now, we need to look at things the other way around and ask what led such a stable culture to develop symbolic thought as we know it. What other elements may have come into play?

We can begin to answer this question by applying the rules of biological evolution to the evolution of culture, and state that culture is based on the survival of ideas that are best suited to the environment in which they are formed. Seen from this viewpoint, the exchange of goods and services would serve the same purpose as that of sex in biological evolution: on one hand, it would increase individual satisfaction, and on the other, it would generate new benefits, meaning more wealth in the case of trade and more offspring in the case of sex. We share the latter benefit with many other species, but the opportunity of increasing wealth through trade is a specific feature of *H. sapiens*. We have seen that this became a possibility when we began to equip ourselves with the earliest tools: someone who had an axe was wealthier than someone who only had two arms to rely on, although this difference could easily be overcome through imitation. We had not yet invented the patent system.

If we refer to the significant increase in (material) wealth that is generated through specialisation and trade, when precisely did these early forms of behaviour start? And where can we find the most ancient evidence of forms of behaviour that could be described as economic, however embryonic?

14.3 First Evidence

Many scholars believe that the first modern human behaviour emerged in Eurasia about 40,000 years ago, coinciding with the first manifestations of cave and movable art. Yet, this form of behaviour was more concerned with artistic rather than economic expression. More and more evidence is now emerging to suggest that our modern behaviour developed much earlier, not in Europe but in Africa.

In the Blombos cave in South Africa, archaeologists found a great number of seashells of the species *Nassarius kraussianus* that were pierced using special awls made out of stone or bone, some blackened by means of elaborate heating techniques or painted with the use of pigments made out of powdered minerals. Many hours of work were required in all these cases. Sometimes, it is clear from wear on the edges of the perforations that the shells were held together by laces. By analysing the traces of wear, it has even been possible to reconstruct how many knots were made to divide and assemble the shells into different styles. These objects do not, however, seem to have had any practical use, and they have also been found at many sites far from the sea. *Nassarius* shells of different species, perforated and painted using similar techniques and dating back 80,000 years or more, have also been found in Morocco, Algeria and Tunisia. Similar finds in the Middle East date back to 100,000 years ago.

There is a reasonable likelihood that the perforated shells were used not only as ornaments, in the form of bracelets and necklaces, but also as a medium of exchange. This means they could have been the first ever form of money. Ironically, a coin with the engraving of a shell was circulating in Ghana up until recently, under the name of *cedi* (meaning cowry shell in Fedi, one of the local languages). Indeed real cowries were formerly used as currency, in the area, alongside coins and gold dust, until 1901.

In any case, were those perforated shells emerging from the deep past really coins? In order to answer this question, we need to consider the requirements that an object must meet before it can be considered a form of currency.

Fig. 14.1 A coin with the engraving of a shell, circulating in Ghana. *Source*: Shutterstock, Copyright vitaliy_73

14.4 The Concept of Money

If we assume, for the moment, that the main function of money is to act as a medium of exchange, the chosen object must first be scarce, otherwise anyone could make money and it would cease to function as an "invariable" measure of the value of the exchanged goods. This would give rise to the problem we now refer to as inflation. If there are no scarce items at hand, any conventional objects could be used by applying some type of complex workmanship to them that is recognisable and difficult to replicate. As an analogy in our own time, we need only think of the signs printed on the first coins or the complex patterns on our paper money, which can only be issued by a special authority holding exclusive rights. In order to stop just anyone from creating money, it should be enough to ensure that more work is applied to its "manufacture" than that necessary to produce the goods it is exchanged for. Alternatively, the work required to create the money should be of "rare quality". Another alternative would be to identify assets that only very gradually increase in availability and use them as money. For example, family herds were used until recently for this purpose, both as standard values and as means of exchange, counting on a herd growth rate curbed by our dietary needs.

Secondly, this object should be portable and suitable for expressing different values, depending on the worth of the goods exchanged. Lastly, it would have to be a non-perishable good, capable of holding its value over time to allow successive exchanges to take place.

If we require these characteristics to apply to whatever we want to function as money, then bracelets and necklaces made up of certain specific shells, if rare or marked with pigment or other methods requiring complex workmanship, could be interpreted as purely ornamental objects or perhaps also as a form of currency. The case in point marks one of the first processes of abstraction in which symbolic thought was expressed: the human representation of a conventional value that was recognised and broadcast within a community.

The end result of this reasoning is the same, and the point is made more powerfully if we look at money in a more modern light, in other words, not merely as a medium of exchange, but also as a set of debit (and therefore also credit) relationships established within a society. According to this interpretation, money is nothing more than transferable credit and the item used as money is largely irrelevant. All we need to find is a system for recording these credit and debit relationships. In this case, as soon as we learned to count, this function could have been fulfilled by knots on a string or regular and repeated notches carved on a board.

We have, indeed, found an ochre tablet deliberately engraved with repeated lines dating back 80,000 years (again in Blombos), and even notches cut into sticks made out of wood, bone or other materials dating back 30,000 years (often referred to as "notched tallies"). Surprisingly enough, these first sticks resemble those belonging to the English Exchequer's collection of tallies, mostly destroyed in 1826, because this accounting system was considered too primitive when compared to the new issue of currency by the Bank of England. In the Middle Ages, all financial relationships between the British Crown, landowners and other creditors were, indeed, recorded by cutting notches on wooden sticks. Yet, so far this is mere speculation.

At the end of our mind's long evolutionary march, most of the currency used in the contemporary world has no physical form. According to statistics published by the Federal Reserve Bank of St Louis and the Bank of England, the intangible component of money (as just described) amounts to 90 % of the currency used in the US and 97 % of that used in Britain, respectively. If archaeologists of the future have to rely on the fossil record, they will understand very little about this aspect of our lives today. Indeed, many of our contemporaries might not understand it either.

14.5 Real Economy

Whether or not money is involved in the real economy, exchanges of goods and services depend on the existence of a surplus of goods and services for exchange.

We have seen that this surplus comes about, in a society, through the division of labour and specialisation into increasingly detailed tasks. The process is facilitated by the evolution of technological progress, which, in our particular case, is reflected by the tools that can be found at various archaeological sites. And technological development is determined by changes in environmental conditions and population trends. What archaeological evidence can we find to support this? We will begin with technologies.

It has been shown that between approximately 100,000 and 70,000 years ago, in various areas of Africa, *sapiens* had developed new methods of working stone, sometimes using pyrotechnology, which enabled state-of-the-art double-sided tools to be produced using heating techniques.

Flint blades (microliths) dating back 71,000 years have been found at the Pinnacle Point site in South Africa. They must have required at least six stages of manufacture (identifying suitable stones, collecting firewood, treating the stones with fire, preparing a stone core, producing blades and finishing them to produce a

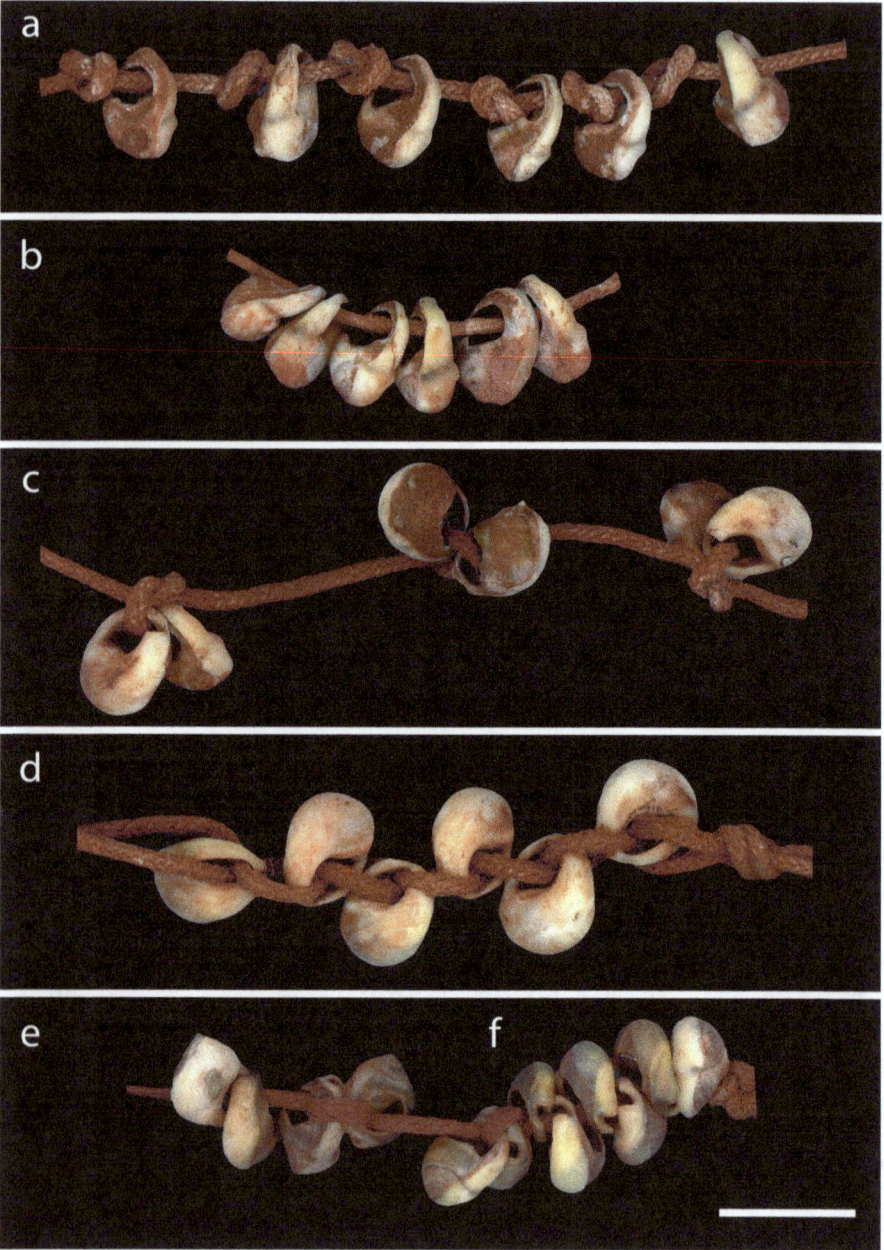

Fig. 14.2 Shells from Blombos: just an ornament or a medium of exchange? *Source*: By kind permission of Francesco d'Errico

variety of final forms). If they were to be incorporated in an arrow or spear, it was necessary to add other materials (feathers, glues) and additional processing stages (bending the wood and fixing techniques).

When we speak of the "stone age", we are referring only to what has been left. Many objects and materials associated with stone-age finds have been lost because they were perishable. We know little about their beauty and function or the craftsmanship required to make them. In any case, the amount of work involved at the end of the entire process may have taken anything from weeks to months and also required the ability to carry out complex operations for very specific purposes and creative techniques, which, in this case, had been handed down for over 10,000 years.

Some suppose that these flint blades were the tips of new deadly throwing spears equipped with a wooden propeller (similar to those still in use by some native peoples of Australia, up until recently). Given the advantage that these weapons would have given to their users, such findings are consistent with the beginning of the arms race associated with territorial defence purposes, as mentioned above when talking about the onset of collective private property along the coasts of South Africa.

14.6 Technology, Population and Climate

What relationships can we envisage, at this point, between the technologies, the demographic variables and the environmental conditions of that age? In principle, dramatic environmental changes work in much the same way as wars: they decimate the population, but also stimulate the intellect and generate rapid technological innovation. This facilitates a greater accumulation of resources, which then translates into a population increase, climate permitting. When the population becomes large enough, the first exchanges of goods and services begin.

In order to set up conditions that are favourable for those exchanges, the surplus generated by technological innovation must initially be greater than that absorbed by the increased population. Later, as a result of increasingly specialised exchanges, this surplus is able to grow even more, promoting a further rise in the population. The result is a virtuous cycle that feeds on itself. But how is this mechanism triggered?

One could argue that neither the environmental crisis alone, nor a sudden genetic mutation that might have helped us develop a complex language, could have been the sole catalysts for such highly accelerated progress to take place. Instead, it must also have required the emergence of a particular form of behaviour: one that was economic and based on the exchange of goods and services.

In order to bring about this cultural revolution, we first had to overcome one enormous obstacle: the inefficiency of the barter system, which must have made it necessary to make repeated exchanges, perhaps even an infinite number, before individuals could come into contact with others selling what they really wanted to buy.

The Blombos shell necklaces, obsidian from volcanic areas, ochre tablets bearing symbolic markings and even the innumerable sophisticated stone blades found throughout the continent could, therefore, be considered not only as objects exchanged in their own right for their intrinsic qualities, perhaps because they had some practical usefulness or some unknown value, but also as intermediaries in other forms of trade. We probably initially progressed from barter pure and simple (though we have no evidence for this) to exchanging all goods and services through intermediate goods, and then ultimately through money. Yet, in the latter case, the wealth generated by the growth of trade depended on an almost exclusively symbolic representation of the value of the goods exchanged, one initially expressed by any form of intermediate asset and then by money itself.

Because these assets were not to be perishable, over the course of time, it was possible to accumulate enormous quantities of money (wealth), partly concentrated in the hands of individuals and partly spent for the benefits of the community on goods and services that could be enjoyed by all. Hence, the emergence of monumental works and the possibility of financing art in all the forms we know up to the present day.

At the dawn of the latest industrial revolution, when wealth accumulated in the form of money (and amplified by the invention of credit letters by the new breed of bankers) was again invested in international trade, another abstract concept was created through the invention of joint-stock companies (the precursors of present-day public limited companies). These took the form of abstract personifications of the investors of that wealth (avatars before their time) that limited any losses to that specific commercial venture, thus protecting the investor's overall wealth. The result was the setting-up of the various West and East Indies Companies, which operated as legal entities separate from their owners, another abstraction.

14.7 Specialisation and Generation of Material Wealth

How did we begin this habit of generating an abstract reality to increase the availability of goods and services and then total wealth? One possible scenario is that the first division of labour took place at the level of gender, with women mainly performing certain specific tasks in exchange for others performed by men. In this case, the situation was not that of an economic specialisation for commercial purposes but of a specialisation whereby goods and services were exchanged (free of charge) based on trust, the bonds of affection, possibly hierarchical relationships and also a certain degree of continuity in time. This applies within any family group and even within a small community. All it takes is a rule of reciprocity or sometimes some exercise of power.

Commercial trade probably began when these elements were removed, for example, when the exchanges were sporadic or took place with outsiders, but the benefits of the division of labour persisted. Such trade was facilitated by the possibility of using any specific object, first as an intermediary and then to settle any indebtedness established during transactions (which somehow had to be tracked). In this case, the trust present in the household was first converted into trust that everyone would acknowledge a conventional method for keeping account of receivables and payables, and subsequently trust that the object used to settle the differences was a transferable credit, with a value that was maintained over time and could therefore be used to accumulate wealth or enable potential future exchange.

When we strip down our most complex contemporary economic behaviour to these principles, it becomes clear that such behaviour first came about in deep time with the emergence of the first division of labour and its application outside the first households. As with all forms of behaviour that have developed over the past 100,000 years, before it could manifest itself to the full, it needed a symbolic tool: something that existed only in our imagination, that could first be represented in the shape of a particular object and then acknowledged as such (through a shared language).

If the suggestions made here are confirmed, we can argue that money was one of the first innovations, together with the arts, to arise out of the process of abstraction manifested in Africa and the Middle East. It must have come about due to adopting economic behaviour as a result of living in wider societies, and it must have been invented to increase the availability of goods and services to meet our needs, both now and in the future.

It is ironic that only a small proportion of the value of each medium of exchange and each debit/credit relationship resides in the object that we have adopted as currency: a little more in rare and shiny metals such as gold and silver, which can also be used for other purposes; much less in any of the coloured paper rectangles covered in symbols that we use for cash; and nothing at all in the electronic transactions so typical of today. Despite this, nearly all of us believe that the money in our pockets or the credit showing in our current account has its own intrinsic value. But we are wrong, because we would not actually be able to do anything with it if no one "recognised" it as money.

14.8 Counter-Evidence

The above arguments can also be used to prove another point: not only to explain what has happened in Africa and Eurasia, but also what has *not* happened elsewhere. For example, they could explain the reason why some *sapiens*, such as those who populated Australia, have had no incentive to use their symbolic thinking to innovate technologically, exchange goods and services, amass surpluses and increase the population. The answer is simple: it has never occurred to them that the available resources might fall short of their needs.

For a population of hunters and gatherers accustomed to moving around, there is no advantage in trade or even in greater specialisation and an excessive increase in the population. You can survive and find all the nourishment (material and spiritual) you need by following the "song-lines", namely musical representations—handed down from generation to generation—of imaginary lines that simultaneously recall creation myths and map certain parts of the territory. That means there is no incentive to trigger the cumulative process described above through symbolism of an economic nature. This may be the reason for the richly spiritual lives that such people typically led immediately prior to our arrival, when they had no concept of private ownership. Such people were not prehistoric; they were mostly pre-economic, as they never had to deal with inescapable "scarcity", thanks to their extraordinary ingenuity in surviving in a harsh environment. In fact, the sole ethnographic evidence of conflict among Australian Aborigines is found in conjunction with concentrated and scarce resources, for example, near those rare and precious billabongs (ponds) that are unable to provide enough drinkable water and nourishment for all.

14.9 Private Accumulation of Wealth

We are left with the mystery of exactly when *Homo sapiens* became *Homo economicus*, namely individuals and then societies capable of generating an increase in wealth that ultimately far exceeded that provided by Mother Nature. We have so far limited ourselves to some bold assumptions, which all require verification, about how we managed to increase our wealth by conducting commercial exchanges sealed by "money". Because there are various definitions of wealth, the times and places at which it may have originated differ according to the definition we adopt.

If, for the sake of argument, we agree to start with a broad (albeit limited) definition of wealth, we suggest that this should be described as the accumulation or the availability of certain resources (tangible or intangible) for the exclusive benefit of some and not others. Such resources would have been defended or conquered through the formation of larger social groups and probably marked the onset of our territoriality. This definition, which focuses on "private" wealth, from which others are excluded, allows us to reconstruct the following wealth accumulation process.

Initially, when larger and larger groups were formed, the first social differences could have arisen because certain individuals demonstrated a greater capacity for abstraction. In this case, wealth would have taken the form of "mental wealth". When restricted to some people only (i.e. shamans), this capacity would have allowed certain individuals to "tell stories", generating the first abstract realities, followed by the first religions and the first codes of conduct (i.e. the first institutions). This would have led to the first inequalities: on the one side, those who told stories and made up the rules, on the other side, those who listened to the stories and observed the rules.

Later, wealth could have consisted of the accumulation of certain durable goods "representing" a certain social status, including ornaments, tools and other token goods, depending on the various cultures. The objects may have been rare or, much more commonly, could have been produced through the input of a certain amount of specialised work. We could define this type of wealth as "relational". It may have been accumulated by certain individuals partly because they were more able or worked harder, but also because of the obligations that they imposed on other group members. In the latter case, it would be based on the benevolence or authority of the former and the gratitude or sense of obligation of the latter. It may, nevertheless, also have been fed by the use of force (and thus based on a degree of authoritarianism). Examples go from enforcing family ties to women and children all the way down to slavery. Over time, this wealth could have been passed down to some descendants, and this would have led to more social inequality.

If this stage were combined with a certain degree of technological progress, the surplus generated may not all have been absorbed by the "wealth deposits" (available only to certain individuals in various chosen forms, for example, spear tips or ornamental objects). Conditions would, therefore, have been ripe for trade, if other groups were willing to dispose of part of their own surpluses.

Only at this point did we achieve what we now know as wealth in the strictest sense, namely the accumulation of resources in the form of money and credit resulting from the production and exchange of goods and services (commercial wealth). This should help us establish which objects were initially chosen to perform the various functions of money, bearing in mind its list of essential characteristics.

If we accept this logical sequence of wealth formation, even if we are uncertain about the time and space at which these phenomena occurred in our deep history, at least we will have more clues for "recognising" and thus interpreting the nature and significance of the objects we find.

The possibility that trade was already taking place in Africa, even before such actions were documented by remains found at various archaeological sites, may seem unlikely, but is not entirely far-fetched. We know that, throughout the period between approximately 100,000 and 10,000 years ago, Africa experienced intense climate fluctuations during which stable, mild periods alternated with environmental catastrophes. This makes it theoretically possible that any wealth accumulated (in various forms) has been generated and destroyed many times, particularly if represented by less durable objects or materials.

The fact that some of the forms of wealth indicated above were already in existence in the deep past, and that these coincided with the first hierarchical relationships and with some division of labour, has been confirmed by a number of finds that we have only touched on in this book. These are associated with certain African sites and the tombs of many of our illustrious ancestors dating back to the ice age in Eurasia. Such discoveries suggest that the symbolism associated with various funerary practices could refer to much more than simply the cult of the dead and other forms of religion—even embracing the political and economic sphere and marking the start of inequality, which may have come about long before the onset of the great civilisations.

Chapter 15
(In)Conclusive Remarks

> *We look on past ages with condescension, as a mere preparation for us... but what if we are only an after-glow of them?*
>
> Farrell (1973, p. 219)

During this short foray into the deep past, we have attempted to tell the story of how we became first humans and then modern humans, starting from the time when other species that came before us embarked on an evolutionary process of which we are merely the most recent expression. To do this, we had to make some partly arbitrary decisions, because there is no limit to how far back we can go in time to shed light on who we are and where we come from.

We decided to tell the story of hominins, beginning at the point at which we assumed an upright position as part of a long process of trial and error, which carried on until circumstances were favourable for making better use of a larger brain. Later on, this led to the advent of symbolic thought and the proliferation of different cultures due to this organ's versatility and plasticity. A number of biological and cognitive innovations first allowed us to adapt to the Earth's growing aridity and then to its increasingly variable climate. This effectively meant going back in time to the moment when our hominin evolutionary line split off from the one that led to the present-day chimpanzees.

We relied on evidence provided by the fossil and archaeological record, as well as on environmental data from the deep past obtained using the latest analytical tools offered by science, but did not need to scour the past for the many nuggets of genetic information that we still carry around with us today. The emerging picture, the surface of which we have only scratched, is already packed with valuable information, much of it very recent and not yet known to the general public. Our attempt to express these results using accessible language leaves us with some issues still obscured, new doubts and a host of more questions. The most important of these is whether there is any essential difference between us and the other species.

Our answer is that we are different because of special cognitive abilities associated with the development of symbolic thought and the establishment of transient cultures.

By summarising all our observations, we have been able to propose a general condition that defines modern humans. This may be described as the ability to create worlds that transcend observed reality and are the outcome of our powers of abstraction. This skill is applied in many fields: from the arts to the sciences; from economics to politics. It is broadcast and multiplied through increasingly complex languages and is expressed in ways that we do not share with any other living species. We have become a social organism, but while naturally occurring social beings, such as beehives and anthills, are governed primarily by biological rules, our species also bears the strong imprint of cultural ties. These govern social behaviour (together with the rules of biology), but often go a step further, as when societies are made up of a great number of individuals who do not know each other. The contribution made by the social sciences to the study of human evolution is particularly useful in explaining this phenomenon.

Where will these underlying traits take us next? We do not have a crystal ball, but we can glimpse at what our future might hold by looking at the subjects of current research projects: genetic engineering, biotechnology, plans for the construction of artificial organs and for a complete mapping of the human brain. Some say we are bypassing natural and sexual selection to plan a new post-modern *H. sapiens*: a freer, healthier, longer-living human who inhabits a more complex and interconnected society based on information and communication technologies (ICT).

With the help of ICT, we believe that we are expanding our individual degrees of freedom by gaining access to a greater amount of information. Whereas such a claim may hold good in general, when backed up by a thorough knowledge of how information is generated and broadcast, one must consider that such technologies also allow for a comprehensive monitoring of the way those individual freedoms are exercised.

Politics as we knew it can thus turn into a deeper control over people; demography can develop into an exploration of what people think; statistics can evolve into Big Data, i.e., the capacity to profile people according to preferences and predict their behaviour. Furthermore, ICT can also be used to create emotions (artificially) and even allow some to market them on a large scale. Lastly, by increasing the complexity of human relations, ICT make a society more vulnerable: the more complex a system—social, in this case—the more exposed it becomes to possible shocks.

To sum up, in principle, more freedom can proceed in parallel with more control; yet, the one that would prevail will not be determined by the available technologies, but rather by the way we choose to use them and the risks we are prepared to take.

We are also left with other problems, which we have highlighted in this book. The unlimited expansion of our capacity for representation has opened a can of worms. This certainly applies to the economy, in which we have created a host of financial and property "bubbles" by assigning imaginary values to the wealth we believe we own. The same applies to our consumption of the planet's resources, which we squander as though they were unlimited. It also applies to the obsessive way we reduce biodiversity and subjugate the animal and plant kingdoms for our

15 (In)Conclusive Remarks

own purposes (which have long been more than mere survival). We also over-equip ourselves for war, given that we could destroy our planet as we know it many times over: the existence and coexistence of many different cultures in conflict with one another does not offer a reassuring picture of a peaceful future.

Our economic, environmental and energetic greed now amounts to a problem that we can no longer ignore. According to the calculations by the Global Footprint Network, a research centre that studies trends in humanity's ecological impact, our planet's ability to replenish resources and absorb the waste we produce only lasts up to August 19 of each year. On August 20, we go into the red, borrowing goods and services from the future, because our ecosystems can no longer regenerate themselves. Plants, animals, clean air and fertile soil: we are using up, at a fast rate, the resources handed down to us by an evolutionary history lasting over 3 billion years. The problem started at least 50,000 years ago, when we began to change ecosystems and bring about the extinction of entire species, leaving many victims in our wake. It continued until the 1970s, when we reached a balance point between renewable resources and consumed resources. Since then, the deficit has been growing. It is estimated that it would now take one and a half planets to produce the resources necessary to sustain humanity's ecological footprint. According to a conservative estimate, it will take three planets before the middle of this century. Time is running out and we must find a solution.

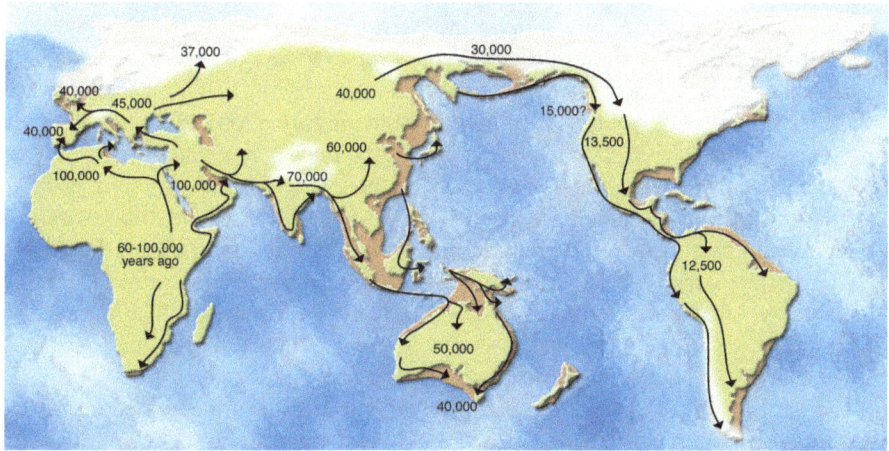

Fig. 15.1 Dispersion of modern *Homo sapiens* based on genetic and archaeological evidence (time scale expressed in years). *Source*: Adapted from D. Palmer, *The Origins of Man*, New Holland Publishers, London 2007

We could begin by reflecting more on the ambivalence of human nature. This means going beyond who we *think* we are. Neither is it useful to examine what we *wish* to be, when we refer to an ideal construct that we think should inspire our actions. The human nature we should really use as our benchmark emerges from the kaleidoscope of behaviour patterns observed during the course of our evolutionary history. These patterns can be summarised in seven simple points, which we offer for our readers' consideration.

1. *Homo sapiens* differ from all other known species due to their desire to dominate nature and bend it to their own ends. The time when they started adopting this attitude is hotly debated and may go very far back in time.
2. The divergence of *Homo* from earlier hominins might go back to when we started to distinguish between *being* and *having*: in other words, between "being endowed", through the forces of natural selection, with some anatomical variation, and "equipping ourselves" with artefacts that stem from our ingenuity in order to increase our chances of survival and growth. This distinction, attributable to our new mental abilities, has had significant repercussions on our cognitive abilities. It has also led to the advent of wealth, which remained limited for millions of years to a few useful items and the control of fire, but then exploded into an increasingly complex series of goods, services and relationships, all based on the division of labour. This brought along with it a progressive increase in inequality, generated and managed through complex symbolic and relatively stable structures, namely cultures.
3. Symbolic thought enabled *Homo sapiens* gradually to create "worlds" that exist only in their mind and govern their actions. By doing so, they were able to organise, grow and progress, increasing their ability to control the environment and the behaviour of single individuals. As a culture became settled, *Homo sapiens* also became *Homo ethicus* and *Homo socialis*.
4. Due to their capacity for abstraction, *Homo sapiens* were also able to understand natural forces that they were unable to *sense* (through the limited senses available to them) and to *perceive* (through their interpretation). Being able to see "other worlds" also turned *Homo sapiens* into *Homo scientificus*.
5. When resources were too scarce to meet their needs, *Homo sapiens* became *Homo economicus*, namely people who, individually, focus only on goods and services to be consumed (which are limited by their ability to afford them), but who collectively seek to increase their availability and appropriate them, no matter whether their ability to regenerate has been exceeded. The interaction between these individual and collective forms of behaviour determines new needs and accelerates the destruction of the available resources.
6. The application of symbolic thought to the various areas mentioned above, as well as others, creates a wide variety of cultures, which are often in conflict with one another, producing different overlaps and clashes between *Homo ethicus*, *Homo socialis*, *Homo scientificus* and *Homo economicus*: these figures are all present in *Homo sapiens*, but to different extents in different cultures. In particular, a recent empirical study has shown that the identification with *Homo economicus* (defined as merely pursuing self-interest) reduces the degree of trust between individuals and, therefore, weakens the benefits that such a figure is supposed to provide to the society as a whole.

7. The excesses to which unbridled symbolic thinking can lead cause us to reflect on the interaction that we wish to see between the different aspects of our nature in the future. In theory, we could decide whether we want to continue to give priority to *Homo economicus*, as we have done over recent centuries in most parts of the world, or whether we wish to look to the future and introduce other criteria for organising ourselves socially and surviving on this planet. For example, we might wish to give precedence to *Homo ethicus* and *Homo socialis*, based on information obtained from *Homo scientificus* and the limits demonstrated by *Homo economicus*. Readers will undoubtedly be able to think of other combinations and imagine the futures they might bring.

One message of this book is that when certain ideas take root and are handed down, they give rise to a culture that can evolve if the new ideas are able to filter through society. The time may have come to imagine a new epoch in which humankind can establish acceptable criteria of sustainability and rein in, as far as possible, its excesses of destruction and representation.

Fig. 15.2 *Source*: Shutterstock, Sunny studio

Further Readings

Books

Arsuaga, J. L. and Martinez, I. 2013, *La Especie Elegida*, Ediciones Planeta, Madrid.
Ayala, J. F. 2015, *¿De dónde vengo? ¿Quién soy? ¿A dónde voy?*, Alianza Editorial, Madrid.
Baggott, J. 2015, *Origins. The Scientific Story of Creation*, Oxford University Press, Oxford.
Begun, D. R. 2015, *The Real Planet of the Apes: A New Story of Human Origins*, Princeton University Press, Princeton.
Cameron, D. W. and Groves, C. P. 2004, *Bones, Stones and Molecules. "Out of Africa" and Human Origins*, Elsevier Academic Press, Oxford.
Clottes J. 2016, *What is Palaeolithic Art. Cave Paintings and the Down of Human Creativity*, The University of Chicago Press, Chicago.
Dunbar, R. 2014, *Human Evolution*, Penguin Books, London.
Fillard, J. P. 2015, *Is Man to Survive Science?* World Scientific, Singapore.
Harari, J. N. 2014, *Sapiens: a Brief History of Humankind*, Harper Collins, New York.
Henke, W. and Tattersall, I. 2015, *Handbook of Paleoanthropology*, Springer-Ferlag, Heidelberg.
Lieberman, D. E. 2014, *The Story of the Human Body: Evolution, Health, and Disease*, Pantheon Books, New York.
Manzi, G. 2013, *Il Grande Racconto dell'Evoluzione Umana*, Il Mulino, Bologna.
Pääbo, S. 2014, *Neanderthal Man: in Search of Lost Genomes*, Basic Books, New York.
Pasternak, C. A. (ed.) 2007, *What Makes Us Human*, Oneworld, Oxford.
Ryan, C and Jethá, C. 2012, *How We Mate, Why We Stray, and What It Means for Modern Relationships*, Harper Perennial.
Stringer, C. 2011, *The Origin of Our Species*, Penguin Group, London.
Tattersall, I. 2015, *The Strange Case of the Rickety Cossack*, Palgrave MacMillan, New York.
Tuniz, C. 2013, *Radioactivity*, Oxford University Press, Oxford.
Tuniz, C., Gillespie, R. and Jones, C. 2009, *The Bone Readers: Atoms, Genes and the Politics of Australia's Deep Past*, Allen & Unwin, Sydney.
Tuniz, C., Manzi, G. and Caramelli, D. 2014, *The Science of Human Origins*, Left Coast Press, Walnut Creek, CA.

Selected Papers and Books for Chapter 1: History, Prehistory and Deep Time

Dennell, R. and Porr, M. (eds.) 2014, *Southern Asia, Australia and the Search for Human Origins*, Cambridge University Press, Cambridge.
Renfrew, C. and Bahn P. (eds.) 2014, *The Cambridge World Prehistory*, Cambridge University Press, Cambridge.

Selected Papers and Books for Chapter 2: Genesis

Bell, E. 2015, *Potentially Biogenic Carbon Preserved 4.1 billion-year-old Zircon*, in "Proceedings of the National Academy of Sciences", 1517557112v1-201517557.
Darwin, C. 1859, *On the Origin of Species by Means of Natural Selection or, the Preservation of Favoured Races in the Straggle for Life*, Murray, London.
Faurby, S. and Svenning, J. C. 2015, *Historic and Prehistoric Human-driven Extinctions have Reshaped Global Mammal Diversity Patterns*, in "Diversity and Distributions", 21, pp. 1155–66
Florio, M. et al. 2015, *Human-Specific Gene ARHGAP11B Promotes Basal Progenitor Amplification and Neocortex Expansion*, in "Science", 347, pp. 1465-70.
Hawking, S. and Mlodinow, L. 2010, *The Grand Design*, Transworld Digital, London.
Hutton, J. 1788, *Theory of the Earth: Or an Investigation of the Laws Observable in the Composition, Dissolution, and Restoration of Land upon the Globe*, in "Transactions of the Royal Society of Edinburgh", 1, Part II, pp. 209-304.
Reeves, H. et al. 2011, *Origins: Cosmos, Earth, and Mankind*, Arcade, New York.

Selected Papers and Books for Chapter 3: The Star Wars Cantina

Armitage, S. J. 2011, *The Southern Route "Out of Africa": Evidence for an Early Expansion of Modern Humans into Arabia*, in "Science", 331, pp. 453-6.
Beleza, S. et al. 2012, *The Timing of Pigmentation Lightening in Europeans*, in "Molecular Biology and Evolution", 30, pp. 24-35
Berger, L. et al. 2015, *Homo Naledi, a New Species of the Genus Homo from the Dinaledi Chamber, South Africa*, in "eLife", 4:e09560.
Brown, P. et al. 2004, *A New Small-Bodied Hominin from the Late Pleistocene of Flores, Indonesia*, in "Nature", 431, pp. 1055-61.
Cann, R. L. et al. 1987, *Mitochondrial DNA and Human Evolution*, in "Nature", 325, pp. 31-6.
Castañeda, I. S. et al. 2009, *Wet Phases in the Sahara/Sahel Region and Human Migration Patterns in North Africa*, in "Proceedings of the Academy of Sciences", 106, pp. 20159-63.
Estairrich, A. and Rosas, A. 2015, *Division of Labor by Sex and Age in Neandertals: an Approach through the Study of Activity-Related Dental Wear*, in "Journal of Human Evolution", 80, 51-63.
Fu, Q. et al. 2015, *An Early Modern Human from Romania with a Recent Neanderthal Ancestor*, in "Nature", 524, pp. 216-219.
Grabowski M. et al., 2015, *Body Mass Estimates of Hominin Fossils and the Evolution of Human Body Size*, in "Journal of Human Evolution" 85, pp. 75-93.
Groucutt, H.S et al. 2015, *Rethinking the Dispersal of Homo sapiens out of Africa*, in "Evolutionary Anthropology", 24, pp. 149–164

Huerta-Sánchez, E. 2014, *Altitude Adaptation in Tibetans Caused by Introgression of Denisovan-like DNA*, in "Nature", 512, pp. 194-197.
Jablonski, N. G. et al. 2010, *Human Skin Pigmentation as an Adaptation to UV Radiation*, in "Proceedings of the Academy of Sciences", 107, pp. 8962-8.
Krause, J. et al. 2010, *The Complete Mitochondrial DNA Genome of an Unknown Hominin from Southern Siberia*, in "Nature", 464, pp. 894-7.
Kuhlwilm, M. et al. 2016, *Ancient Gene Flow from Early Modern Humans into Eastern Neanderthals*, in "Nature", doi:10.1038/nature16544
Morwood, M. J. et al. 2005, *Further Evidence for Small-Bodied Hominins from the Late Pleistocene of Flores, Indonesia*, in "Nature" 437, pp. 1012-7.
Pearce, E. et al. 2013, *New Insights into Differences in Brain Organization between Neanderthals and Anatomically Modern Humans*, in "Proceedings of the Royal Society B: Biological Sciences", 280, pp. 1-7.
Peresani, M. et al. 2011, *Late Neandertals and the Intentional Removal of Feathers as Evidenced from Bird Bone Taphonomy at Fumane Cave 44 ky B.P., Italy*, in "Proceedings of the Academy of Sciences", 108, pp. 3888-93.
Pike, A. W. G. et al. 2012, *U-Series Dating of Paleolithic Art in 11 Caves in Spain*, in "Science", 336, pp. 1409-13
Raghavan, M. 2014, *Upper Palaeolithic Siberian genome reveals dual ancestry of Native Americans*, in "Nature", 505, pp. 87-91.
Rasmussen, M. et al. 2011, *An Aboriginal Australian Genome Reveals Separate Human Dispersals into Asia*, in "Science", 334, pp. 94-8.
Reich, D. et al. 2010, *Genetic History of an Archaic Hominin Group from Denisova Cave in Siberia*, in "Nature", 468, pp. 1053-60.
Rodríguez-Vidal, J. et al. 2014, *A Rock Engraving Made by Neanderthals in Gibraltar*, in "Proceedings of the National Academy of Sciences", 111, pp. 13301-6.
Romandini, N. et al. 2014, *Convergent Evidence of Eagle Talons Used by Late Neanderthals in Europe: A Further Assessment on Symbolism*, in "Proceedings of the Academy of Sciences", 9 (7), e101278.
Sawyer, S. et al. 2015, *Nuclear and Mitochondrial DNA Sequences from Two Denisovan Individuals*, in "Proceedings of the National academy of Sciences", doi/10.1073/pnas.1519905112
Schwartz J. H. and Tattersall, I. 2015, *Defining the Genus Homo*, in "Science", 349, pp. 931-932.
Soressi, M. et al. 2013, *Neanderthals Made the First Specialized Bone Tools in Europe*, in "Proceedings of the National Academy of Sciences", 110, pp. 14186-90.
Sutikna, T. et al. 2016, *Revised Stratigraphy and Chronology for Homo floresiensis at Liang Bua in Indonesia*, in "Nature", doi:10.1038/nature17179
Tuniz, C. et al. 2012, *Did Neanderthal Play Music? X-ray Computed Micro-Tomography of the Divje Babe "flute"*, in "Archaeometry", 54, pp. 581-90.
van den Berg, G. D. et al. 2016, *Earliest hominin occupation of Sulawesi, Indonesia*, in "Nature", 529, pp. 208-11

Selected Papers and Books for Chapter 4: The Apes and Us

Haile-Selassie, Y. 2015, *New Species from Ethiopia further Expands Middle Pliocene Hominin Diversity*, in "Nature", 521, pp. 483-488.
Patterson, N. et al. 2006, *Genetic Evidence for Complex Speciation of Humans and Chimpanzees*, in "Nature", 441, pp. 1103-08.
Raymo, M. E. and Huybers, P. 2008, *Unlocking the Mysteries of the Ice Ages*, in "Nature", 451, pp. 284-5.
Walter, R. C. 1994, *Age of Lucy and the First Family: Single-crystal 40Ar/39Ar Dating of the Denen Dora and Lower Kada Hadar Members of the Hadar Formation, Ethiopia*, in "Geology", 22, pp. 6-10.

White, T. D. et al. 2009, *Ardipithecus ramidus and the Paleobiology of Early Hominids*, in "Science", 326, pp. 75-86.
Zimmer, C. 2015, *When Darwin Met Another Ape*, in "National Geographic", April 21.
White, T. D. et al, 2015, *Neither Chimpanzee nor Human, Ardipithecus Reveals the Surprising Ancestry of Both*, in" in "Proceedings of the National Academy of Sciences", 112, pp. 4877-84.
Zollikofer, C. P. E. et al. 2005, *Virtual Cranial Reconstruction of Sahelanthropus tchadensis*, in "Nature", 434, pp. 755-9.

Selected Papers and Books for Chapter 5: "The Quest for Fire"

Almécija, S. et al. 2015, *The Evolution of Human and Ape Hand Proportions*, in "Nature Communications", 6, doi: 10.1038/ncomms8717.
Benazzi, S. et al. 2015, *The Makers of the Protoaurignacian and Implications for Neandertal Extinction*, in "Science", 348, pp. 793-796.
Briggs, A.W. et al. 2009, *Targeted Retrieval and Analysis of Five Neandertal mtDNA Genomes*, in "Science", 325, pp. 318-321.
García-Diez, M. et al. 2013, *Uranium Series Dating Reveals a Long Sequence of Rock Art at Altamira Cave (Santillana del Mar, Cantabria)*, in "Journal of Archaeological Science", 40, 4098-4106.
Green, R. E. et al. 2008, *A Complete Neanderthal Mitochondrial Genome Sequence Determined by High-Throughout Sequencing*, in "Cell", 134, pp. 416-26.
Green, R. E. et al. 2010, *A Draft Sequence of the Neanderthal Genome*, in "Science", 328, pp. 710-22.
Harmand, S. et al. 2015, *3.3-Milion-Year-Old Stone Tools from Lomekwi 3, West Turkana, Kenya*, in "Nature", 521, pp. 310-15.
Higham, T. et al. 2014, *The Timing and Spatiotemporal Patterning of Neanderthal Disappearance*, in "Nature", 512, pp. 306-9.
Maslin, M. M. et al. 2015, *A Synthesis of the Theories and Concepts of Early Human Evolution*, in "Philosophical Transaction of the Royal Society of London – Series B: Biological Sciences", B370: 20140064.
Mellars, P. and French, J. C. 2011, *Tenfold Population Increase in Western Europe at the Neanderthal-to-Modern Human Transition*, in "Science", 333, pp. 623-7.
Miller, G. et al. 2015 *Human Predation Contributed to the Extinction of the Australian Megafaunal Bird Genyornis newtoni ~47 ka*, in "Nature Communications, DOI: 10.1038/ncomms10496.
Potts, R. and Faith, J. T, 2015, *Alternating Low Climate Variability: The Context of Natural Selection and Speciation in Plio-Pleistocene Hominin Evolution*, in "Journal of Human evolution", http://dx.doi.org/10.1016/j.jhevol.2015.06.014
Prüfer, K. et al. 2014, *The Complete Genome Sequence of a Neanderthal from the Altai Mountains*, in "Nature", 505, pp. 43-9.
Roach, N. T. 2013, *Elastic Energy Storage in the Shoulder and the Evolution of High-Speed Throwing in Homo*, in "Nature", 498, pp. 483-486.
Roberts, L. G. et al. 2001, *New Ages for the Last Australian Megafauna: Continent- Wide Extinction about 46,000 Years ago*, in "Science", 292, pp. 1888-92.
Roberts, L. G. et al. 2016, *Climate Change not to Blame for late Quaternary megafauna extinctions in Australia*, in "Nature Communications", · Doi: 10.1038/ncomms10511
Schröder, I. et al. 2014, *Characterizing the Evolutionary Path(s) to Early Homo*, in "PLoS ONE", 9, 12, e114307.
Villmoare, B. et al. 2015, *Early Homo at 2.8 Ma from Ledi-Geraru, Afar, Ethiopia*, in "Science", 347, 1352-55.

Selected Papers and Books for Chapter 6: The Naked Ape

Kittler, R. et al. 2003, *Molecular Evolution of Pedinculus Humanus and the Origin of Clothing*, in "Current Biology", 13, pp. 1414-7.
Sutou, S. 2012, *Hairless Mutation: A Driving Force of Humanization from a Human-Ape Common Ancestor by Enforcing Upright Walking while Holding a Baby with Both Hands*, in "Genes to Cells", 17, pp. 264-72.
Toups, M. A. et al. 2011, *Origin of Clothing Lice Indicates Early Clothing Use by Anatomically Modern Humans in Africa*, in "Molecular Biology and Evolution", 28, pp. 29-32.
Trinkaus, E. and Shang H. 2008, *Anatomical Evidence for the Antiquity of Human Footwear: Tianyuan and Sunghir*, in "Journal of Archaeological Science", 35, pp. 1928-33.
Turk, J. et al. 2015, *Hair Imprints in Pleistocene Cave Sediments and the Use of X-ray Micro-computed Tomography for their Reconstruction*, in "Facies", 61, pp. 2-11.

Selected Papers and Books for Chapter 7: Lucy and the Other Ladies

Alemseged, Z. et al. 2006, *A Juvenile Early Hominin Skeleton from Dikika, Ethiopia*, in "Nature", 443, pp. 296-301.
Austin, C. et al. 2013, *Barium Distributions in Teeth Reveal Early-Life Dietary Transitions in Primates*, in "Nature", 498, pp. 216-9.
Azéma, M. and Rivère, F. 2012, *Animation in Palaeolithic Art: A Pre-Echo of Cinema*, in "Antiquity", 86, pp. 316-24.
D'Anastasio, R. et al. 2013, *Micro-Biomechanics of the Kebara 2 Hyoid and Its Implications for Speech in Neanderthals*, in "PLoS ONE", 8, 12, e82261.
Dean, M. C. and Elamin, F. 2014, *Parturition Lines in Modern Human Wisdom Tooth Roots: Do They Exist, Can They Be Characterized and Are They Useful for Retrospective Determination of Age First Reproduction and/or Inter-Birth Intervals?*, in "Annals of Human Biology", 41, pp. 358-67.
Falk, D. 2004, *Prelinguistic Evolution in Early Hominins: Whence Motherese?*, in "Behavioural and Brain Sciences", 27, pp. 491-503.
Gunz, P. et al. 2010, *Brain Development After Birth Differs Between Neanderthals and Modern Humans*, in "Current Biology", 20, pp. R921-2
McPherron, S. P. et al. 2010, *Evidence for Stone-tool-assisted Consumption of Animal Tissues Before 3.39 Million Years Ago at Dikika, Ethiopia*, in "Nature", 466, pp. 857-60.
Morgan, E. 1972, *The Descent of Woman*, Souvenir Press, London.
Rosenberg, K. and Trevathan, W. 1995, *Bipedalism and Human Birth: The Obstetrical Dilemma Revisited*, in "Evolutionary Anthropology", 4, pp. 161-8.
Simpson, S. W. et al. 2008, *A Female Homo Erectus Pelvis from Gona, Ethiopia*, in "Science", 322, pp. 1089-92.

Selected Papers and Books for Chapter 8: Menus of the Past

Antón, S. C. et al. 2014, *Evolution of Early Homo: An Integrated Biological Perspective*, in "Science", 345, pp. 1-13.
Benedetti, F. 2012, *L'effetto placebo*, Carocci, Roma.

Cerling, T. E. et al. 2011, *Diet of Paranthropus Boisei in the Early Pleistocene of East Africa*, in "Proceedings of the National Academy of Sciences", 108, pp. 9337-41.
Fiorenza, l. et al. 2015, *To Meat or Not to Meat? New Perspectives on Neanderthal Ecology*, in "Yearbook of Physical Anthropology", 156, pp. 43-71.
Hardi, K. et al. 2012, *Neanderthal Medics? Evidence for Food, Cooking, and Medicinal Plants Entrapped in Dental Calculus*, in "Naturwissenschaften", 99, pp. 617-26.
Henry, A. G. et al. 2010, *Microfossils in Calculus Demonstrate Consumption of Plants and Cooked Foods in Neanderthal Diets (Shanidar III, Iraq; Spy I and II, Belgium)*, in "Proceedings of the National Academy of Sciences", 108, pp. 486-91.
Henry, A. G. et al. 2012, *The Diet of Australopithecus Sediba*, in "Nature", 487, pp. 90-3.
Jones, R. S. 2015, *Space Diet: Daily Mealworm (Tenebrio molitor) Harvest on a Multigenerational Spaceship*, in "Journal of Interdisciplinary Science Topics", 4, pp. 1-4.
Lalueza-Fox, C. et al. 2009, *Bitter Taste Perception in Neanderthals through the Analysis of the TAS2R38 Gene*, in "Biology Letters", 5, pp. 809-11.
Ledogar, J.A. et al. 2016, *Mechanical evidence that Australopithecus sediba was limited in its ability to eat hard foods*, in "Nature Communications", doi: 10.1038/ncomms10596
Lieberman, D. E. et al. 2016, *Impact of Meat and Lower Palaeolithic Food Processing Techniques on Chewing in Humans*, in "Nature", doi:10.1038/nature16990
Logan, A. C. et al. 2015, *Natural Environments, Ancestral Diets, and Microbial Ecology: is there a Modern "Paleo-deficit disorder"?* Part I and II, in "Journal of Physiological Anthropology", 34, doi 10.1186/s40101-015-0041-y and doi 10.1186/s40101-014-0040-4.
Marean, C. W. 2015, *An Evolutionary Anthropological Perspective on Modern Human Origins*, in "Annual Review of Anthropology", 44.
Mirazón Lahr, M. et al. 2016, *Inter-group Violence Among Early Holocene Hunter-Gatherers of West Turkana, Kenya*, in "Nature", doi:10.1038/nature16477
Schwitalla, A. W. et al. 2014, *Violence Among Foragers: The Bioarchaeological Record from Central California*, in "Journal of Anthropological Archaeology", 33, pp. 66-83.
Sistiaga, A. et al. 2014, *The Neanderthal Meal: A New Perspective Using Faecal Biomarker*, in "PLoS ONE", 9(6): e101045. doi:10. 1371/journal.pone.0101045.
Smith, G. M. 2015, *Neanderthal Megafaunal Exploitation in Western Europe and Its Dietary Implications: A Contextual Reassessment of La Cotte de St Brelade (Jersey)*, in "Journal of Human Evolution", 78, pp. 181-201.
Sponheimer, M. and Lee-Thorp J. 1999, *Isotopic Evidence for the Diet of an Early Hominid, Australopithecus Africanus*, in "Science", 283, pp. 368-70.
Spoor, F. et al. 2015, *Reconstructed Homo Habilis Type OH 7 Suggests Deep-Rooted Species Diversity in Early Homo*, in "Nature", 519, pp. 83–86.
Van Huis, A. et al. 2013, *Edible Insects: Future Prospects for Food and Feed Security*, in "www.fao.org", Paper 171.

Selected Papers and Books for Chapter 9: Ancient IIIs and Ancient Remedies

Bernardini, F. et al. 2012, *Beeswax as Dental Filling on a Neolithic Human Tooth*, in "PLoS ONE", 7, 9, e44904.
Condemi, S. et al. 2013, *Possible Interbreeding in Late Italian Neanderthals? New Data from the Mezzena Jaw (Monti Lessini, Verona, Italy)*, in "PLoS ONE", 8, 3, e59781.
Coppa, A. et al. 2006, *Palaeontology: Early Neolithic Tradition of Dentistry*, in "Nature", 440, pp. 755-6

Gorjanović-Kramberger G. 1906, *Der diluviale Mensch von Krapina in Kroatien. Ein Beitrag zur Palaöanthropologie*. in Walkhoff, O, (ed), "Studienüber die Entwicklungsmechanik des Primatenskelletes", Volume II. Wiesbaden: Kreidel, 59–277.

Kirsten, I. B. et al. 2014, *Pre-Columbian Mycobacterial Genomes Reveal Seals as a Source of New World Human Tuberculosis*, in "Nature", 514, pp. 494-497.

Lebel, S. and Trinkaus, E. 2001, *A Carious Neanderthal Molar from the Bau de l'Aubésier, Vaucluse, France*, in "Journal of Archaeological Science", 29, pp. 555-7.

Lozano, M. et al. 2013, *Toothpicking and Periodontal Disease in a Neanderthal Specimen from Cova Forada Site (Valencia, Spain)*, in "PLoS ONE", 8, e76852.

Monge, J. et al 1913, *Fibrous Dysplasia in a 120,000+ Year Old Neandertal from Krapina, Croatia*, in "PLoS ONE", 8(6): e64539. doi:10.1371/journal.pone.0064539.

Oxilia, G. et al. 2015, *Earliest Evidence of Dental Caries Manipulation in the Late Upper Palaeolithic*, in "Nature. Scientific Reports", doi: 10.1038/srep12150.

Sankararaman, S. et al. 2014, *The Genomic Landscape of Neanderthal Ancestry in Present-Day Humans*, in "Nature", 507, pp. 354-7.

Simonti, C. N. et al. 2016, *The Phenotypic Legacy of Admixture Between Modern Humans and Neandertals*, in "Science", 351, pp. 737-41.

Trinkaus, E. 1983, *The Shanidar Neandertals*, Academic press, New York.

Zanolli C. and Mazurier A. 2013, *Endostructural Characterization of the H. Heidelbergensis Dental Remains from the Early Middle Pleistocene Site of Tighenif, Algeria*, in "Comptes Rendus Palevol", 12, pp. 293-304.

Selected Papers and Books for Chapter 10: The Hominin Lifestyle

Appenzeller, T. 2013, *Neanderthal Culture: Old Masters*, in "Nature", 497, pp. 302-4.

Carrigan, M. A. 2015, *Hominids Adapted to Metabolize Ethanol Long before Human-Directed Fermentation*, in "Proceedings of the National Academy of Sciences", 112, pp. 458-63.

Caspari R. and Sang-Hee, L. 2004, *Older Age Becomes Common Late in Human Evolution*, in "Proceedings of the National Academy of Sciences", 101, pp. 10895-900.

Henshilwood, C. S. et al. 2002, *Emergence of Modern Human Behaviour: Middle Stone Age Engravings from South Africa*, in "Science", 295, pp. 1278-80.

Joordens, J. C. A. et al. 2015, *Homo Erectus at Trinil on Java used Shells for Tool Production and Engraving*, in "Nature", 518, pp. 228-31.

Martin, R. 2013, *How We Do It: The Evolution and Future of Human Reproduction*, Basic Books, New York.

Mills, K. L. et al. 2014, *The Developmental Mismatch in Structural Brain Maturation During Adolescence*, in "Developmental Neuroscience", 36, pp. 147-60.

Purzycki, B. G. et al., 2016, *Moralistic Gods, Supernatural Punishment and the Expansion of Human Sociality*, in "Nature", doi:10.1038/nature16980.

Shanley, D. P. et al. 2007, *Testing Evolutionary Theories of Menopause*, in "Proceeding of the Royal Society B: Biological Sciences", doi:10.1098/rspb.2007.1028.

Smith, T. M. et al. 2015, *Dental Ontogeny in Pliocene and Early Pleistocene Hominins*, in "PLoS ONE", doi:10.1371/journal.pone.0118118.

Smith, T. M. et al. 2010, *Dental Evidence for Ontogenetic Differences Between Modern Humans and Neanderthals*, in "Proceedings of the National Academy of Sciences", 107, pp. 20923–28.

Tuniz, C. et al. 2012, *A New Assessment of the Neanderthal Child Mandible from Molare, SW Italy, Using X-Ray Microtomography*, in "American Journal of Physical Anthropology", 54, 92 doi:10.1002/ajpa.22033.

Selected Papers and Books for Chapter 11: The Dear Departed of the Pleistocene

Einwögerer, T. et al. 2006, *Upper Palaeolithic Infant Burials*, in "Nature", 444, p. 285.
Formicola, V. and Buzhilova, A. P. 2004, *Double Child Burial from Sunghir (Russia): Pathology and Inferences for Upper Palaeolithic Funerary Practices*, in "American Journal of Physical Anthropology", 124, pp. 189-98.
Pettit, P. B. et al. 2003, *The Gravettian Burial Known as the Prince ("Il Principe"): New Evidence for His Age and Diet*, in "Antiquity", 77, pp. 15-9.
Rendu, W. et al. 2014, *Evidence Supporting an Intentional Neanderthal Burial at La Chappelle-aux-Saints*, in "Proceedings of the National Academy of Sciences", 111, pp. 81-6.
Riel-Salvatore, J. and Gravel-Miguel, C. 2013, *Upper Palaeolithic Mortuary Practice in Eurasia: A Critical Look at the Burial Record*, in Tarlow, S. and Nilsson Stutz, L. (eds.) "The Oxford Handbook of the Archaeology of Death and Burial", Oxford University Press, Oxford, pp. 303-46.
Vanhaeren, M. and d'Errico, F. 2001, *Personal Ornaments from the La Madeleine Child (Peyrony Excavations): An Insight into Upper Palaeolithic Childhood*, in "Revue d'Archeoléologie Préhistorique", 13, pp. 201-40.

Selected Papers and Books for Chapter 12: Brain Readers

Berger, L. R. 2010, *Australopithecus Sediba: A New Species of Homo-Like Australopith from South Africa*, in "Science", 328, pp. 195-204.
Bruner, E. (ed.) 2014, *Human Paleoneurology*, Springer, New York.
Bruner, E. et al. 2014, *Extended Mind and Visuo-Spatial Integration: Three Hands for the Neanderthal Lineage*, in "Journal of Anthropological Sciences", 92, pp. 273-80.
Bruner, E. et al. 2016, *Evidence for Expansion of the Precuneus in Human Evolution*, in "Brain Structure and Function", doi:10.1007/s00429-015-1172-y.
Catapano, R. et al. 2014, *Capuchin Monkeys Do not Show Human-Like Pricing Effects*, in "Frontiers in Psychology", 5, pp. 1-12.
Cela-Conde, C. J. et al. 2014, *In the Light of Evolution: The Human Mental Machinery*, The National Academies Press, Washington (DC).
Dunbar, R. I. M. 1995, *Neocortex Size and Group Size in Primates: A Test of the Hypothesis*, in "Journal of Human Evolution", 28, pp. 287-96.
Falk, D. 2012, *Hominin Paleoneurology: Where Are We Now?*, in "Progress in Brain research", 195, pp. 255-72
Gould, J. G. and Vrba, E. S 1982, *Exaptation - A Missing Term in the Science Form*, in "Paleobiology", 8, pp. 4-15.
Gowlett, J. et al. 2012, *Human Evolution and the Archaeology of the Social Brain*, in "Current Anthropology", 53, pp. 693-722.
Powell J. et al. 2012, *Orbital Prefrontal Cortex Volume Predicts Social Network Size: An Imaging Study of Individual Differences in Humans*, in "Proceeding of the Royal Society B: Biological Sciences", 279, pp. 2157-62.
Rizzolati, G. and Craighero, L. 2004, *The Mirror-Neuron System*, in "Annual Review of Neuroscience", 27, pp. 169-92.
Stout, D. and Khreisheh, N. 2015, *Skill Learning and Human Brain Evolution: An Experimental Approach*, in "Cambridge Archaeological Journal", 25, 867-75.

Selected Papers and Books for Chapter 13: "All Power to Imagination"

Gómez-Robles A. et al. 2015, *Relaxed Renetic Control of Cortical Organization in Human Brains Compared with Chimpanzees*, in "PLoS ONE", doi: 10.1073/pnas.1512646112.
Greenfield, S. 2015, *Mind Change. How Digital Technologies are Leaving their Mark on our Brain*, Random House, London.
Henshilwood, C. S. et al. 2004, *Middle Stone Age Shell Beads from South Africa*, in "Science", 304, p. 404.
Henshilwood, C. S. et al. 2009, *Engraved Ochre from the Middle Stone Age Levels at Blombos Cave, South Africa*, in "Journal of Human Evolution", 57, pp. 27-47.
Hurley, D. 2014, *Smarter. The New Science of Building Brain Power*, Penguin Group, London.
Mar, A.M. 2011, *The Neural Bases of Social Cognition and Story Comprehension*, in "Annual Review of Psychology", 62, pp. 103-134.
Mcbrearty, S. and Brook, A. S. 2000, *The Revolution that Wasn't: a New Interpretation of the Origin of Modern Human Behavior*, in "Journal of Human Evolution", 39, pp. 453-563.
Meshberger, F. L. 1990, *An Interpretation of Michelangelo's Creation of Adam Based on Neuroanatomy*, in "Journal of American Medical Association", 264, pp. 1837-1841.
Miller, G. 2000, *The Mating Mind: How Sexual Choice Shaped the Evolution of Human Nature*, Doubleday, New York.
Powell, A. et al. 2009, *Late Pleistocene Demography and the Appearance of Modern Human Behaviour*, in "Science", 324, pp. 1298-301.

Selected Papers and Books for Chapter 14: Primordial Economy

Brown, K. S. et al. 2012, *An Early and Enduring Advanced Technology Originating 71,000 Years ago in South Africa*, in "Nature", 491, pp. 590-3.
d'Errico et al. 2012, *Early Evidence of San Material Culture Represented by Organic Artifacts from Border Cave, South Africa*, in "Proceedings of the National Academy of Sciences", 109, pp. 13214-19.
d'Errico, F. and Vanhaeren, M. 2014, *La Richesse, Question de Définition*, in "Pour la Science", 445, 10-11.
Henshilwood, C. S., et al. 2009, *Engraved Ochres from the Middle Stone Age Levels at Blombos Cave, South Africa*. in "Journal of Human Evolution", 57, 27-47.
Jacobs, Z. et al. 2008, *Ages for the Middle Stone Age of Southern Africa: Implications for Human Behaviour and Dispersal* in "Science", 322, pp. 733-5.
Martin, F. 2014, *Money: An Unauthorized Biography*, Knopf, New York.
Mcbrearty, S. 2012, *Palaeoanthropology: Sharpening the Mind*, in "Nature", 491, pp. 531-2.
Ofek, H. 2001, *Second Nature: Economic Origins of Human Evolution*, Cambridge University Press, Cambridge.
Ridley, M. W. 2014, *The Rational Optimist*, Fourth Estate, London.
Tiberi Vipraio, P. 1999, *Dal Mercantilismo alla Globalizzazione. Lo Sviluppo Industriale Trainato dalle Esportazioni*, il Mulino, Bologna.
Vanhaeren, M. and d'Errico, F. 2005, *Grave Goods from the Saint-Germain-la-Rivière Burial: Evidence for Social Inequality in the Upper Palaeolithic*, in "Journal of Anthropological Archaeology", 24, pp. 117-34.
Vanhaeren, M. et al. 2013, *Thinking Strings: Additional Evidence for Personal Ornament Use in the Middle Stone Age of Blombos Cave, South Africa*, in "Journal of Human Evolution", 64, pp. 500-17.

Selected Papers and Books for Chapter 15: (In)Conclusive Remarks

Bergoglio, F. 2015, *Encyclical Letter Laudato si'. On care for our common home.* http://w2.vatican.va/content/francesco/en/encyclicals/documents/papa-francesco_20150524_enciclica-laudato-si.html
Byung-Chul Han 2014, *Psicopolítica*, Herder Editorial, Barcellona.
Farrell, J. C. 1973, *The Siege of Krishnapur*, George Weidenfeld and Nicolson, London.
Wilson, E. O. 2012, *The Social Conquest of Earth*, Liveright Publishing Corporation, New York.
   http://www.footprintnetwork.org/en/index.php/GFN/page/world_footprint.
Tiberi Vipraio, P. (ed) 1990, *Etica ed Economia*, Cedam, Padova
Xin Z., Liu G. 2013, *Homo Economicus Belief Inhibits Trust*, in "PLoS ONE", 8(10), e76671, doi:10.1371/journal.pone.0076671

GPSR Compliance

The European Union's (EU) General Product Safety Regulation (GPSR) is a set of rules that requires consumer products to be safe and our obligations to ensure this.

If you have any concerns about our products, you can contact us on

ProductSafety@springernature.com

In case Publisher is established outside the EU, the EU authorized representative is:

Springer Nature Customer Service Center GmbH
Europaplatz 3
69115 Heidelberg, Germany

www.ingramcontent.com/pod-product-compliance
Lightning Source LLC
LaVergne TN
LVHW010343260326
834688LV00036B/853